C000126739

New Directions in Philosophy and Cognitive Science

This series brings together work that takes cognitive science in new directions. Hitherto, philosophical reflection on cognitive science or perhaps better, philosophical contribution to the interdisciplinary field that is cognitive science has for the most part come from philosophers with a commitment to a representationalist model of the mind. However, as cognitive science continues to make advances, especially in its neuroscience and robotics aspects, there is growing discontent with the representationalism of traditional philosophical interpretations of cognition. Cognitive scientists and philosophers have turned to a variety of sources phenomenology and dynamic systems theory foremost among them to date to rethink cognition as the direction of the action of an embodied and affectively attuned organism embedded in its social world, a stance that sees representation as only one tool of cognition, and a derived one at that. To foster this growing interest in rethinking traditional philosophical notions of cognition using phenomenology, dynamic systems theory, and perhaps other approaches yet to be identified we dedicate this series to "New Directions in Philosophy and Cognitive Science."

More information about this series at
http://www.palgrave.com/gp/series/14744

Majid Davoody Beni

Structuring the Self

palgrave
macmillan

Majid Davoody Beni
Department of History, Philosophy, and Religious Studies
Nazarbayev University
Nur-Sultan city, Kazakhstan

The Amirkabir University of Technology
Tehran, Iran

New Directions in Philosophy and Cognitive Science
ISBN 978-3-030-31101-8   ISBN 978-3-030-31102-5   (eBook)
https://doi.org/10.1007/978-3-030-31102-5

© The Editor(s) (if applicable) and The Author(s), under exclusive licence to Springer Nature Switzerland AG 2019
This work is subject to copyright. All rights are solely and exclusively licensed by the Publisher, whether the whole or part of the material is concerned, specifically the rights of translation, reprinting, reuse of illustrations, recitation, broadcasting, reproduction on microfilms or in any other physical way, and transmission or information storage and retrieval, electronic adaptation, computer software, or by similar or dissimilar methodology now known or hereafter developed.
The use of general descriptive names, registered names, trademarks, service marks, etc. in this publication does not imply, even in the absence of a specific statement, that such names are exempt from the relevant protective laws and regulations and therefore free for general use.
The publisher, the authors and the editors are safe to assume that the advice and information in this book are believed to be true and accurate at the date of publication. Neither the publisher nor the authors or the editors give a warranty, express or implied, with respect to the material contained herein or for any errors or omissions that may have been made. The publisher remains neutral with regard to jurisdictional claims in published maps and institutional affiliations.

Cover illustration: World History Archive / Alamy Stock Photo

This Palgrave Macmillan imprint is published by the registered company Springer Nature Switzerland AG.
The registered company address is: Gewerbestrasse 11, 6330 Cham, Switzerland

To my loved ones, Maaedeh, Zohre, Maryam, Ebrahim, and Milad
(and Maaedeh's folk too :))

Preface

I have been busy wresting some information out of my self—about its essence, identity, properties, and modes—for the past thirty-some years. The few things that I have learned hardly make me knowledgeable about the subject, which seems to be elusive as well as intriguing. Even so, a little structured knowledge is not so bad as a lot of scattered ignorance, and giving structure to things that one learns can always pave the way to further wisdom. A great deal of the little that I know about the subject is garnered through pondering the works of great philosophers of the days of lore and recent scientific enterprises as well as introspection. The book does not aim to present a comprehensive philosophical theory that covers everything that we desire to know about the self. Instead, it seeks to give structure to a part of what we already know about the aspects and elements of the self from a scientific and metaphysical point of view. That is to say, in this book I will focus on some specific neuro-computational and informational theories of aspects of the self to flesh out a philosophical alternative to the traditional metaphysics of selfhood. My enterprise is inspired by a flourishing theory of the contemporary philosophy of science.

Structural realism (SR), as being advocated by John Worrall, Steven French, and James Ladyman among others, is a successful theory of the philosophy of science. It is a modest form of scientific realism, and it aims to account for scientific progress and continuity in the history of science on the basis of structural relations, rather than content. It emphasises the unifying role of underpinning structures and underscores the role of commonality in the face of theoretical changes and theoretical diversity that wreak havoc with the realist (epistemological and ontological) commitments of scientific theories. SR defends scientific realism in the face of metaphysical underdetermination in physics. In this book, I draw attention to the divergence of the scientifically informed conception of the self from the orthodox substantivalist picture that is usually associated with the Cartesian tradition. I also draw attention to the diversity that rules over scientific theories of aspects and elements of selfhood. From the observation of instances of divergence and diversity, I conclude that there is a state of underdetermination of metaphysics of selfhood by diverse scientific theories that represent the self and its various aspects. After stating my formulation of metaphysical underdetermination of the self, I argue that a structural realist theory of the self (SRS) presents a strategy for defending a modest form of realism about the self. It does not endorse full-fledged realism about all aspects of the self. It defends a modest version of the self about the basic structure of the self that can be specified in terms of embodied informational structures, or structures realised by mechanisms of information processing in the brain and environment.

Substantivalism, or the view that portrays the self as a classical substance that retains its identity over time and is the bearer of properties, has been the orthodox theory of the self for a long time. Recent scientific (or scientifically informed) theories of the self, for example, Thomas Metzinger's eliminativism or Shaun Gallagher's pattern theory, defy substantivalism. However, they do not provide a well-posed metaphysical alternative to substantivalism. Actually, the domain of cognitive science abounds with a form of theoretical diversity that may distract from the appearance of a viable unified alternative to the orthodox picture. This is because theories of Metzinger and Gallagher, as well as a few other recent theories that I will review in this book, pull in different directions, to the effect that they distract from a unifying ontological account of the self. SRS aims to provide a viable alternative to substantivalism. This alternative is informed by scientific accounts of different aspects of the self, but uses a new metaphysical approach (i.e., ontic structural realism) to staple these theoretical accounts into a unified structural ontology. It constructs its ontological account of the self on the basis of the common underpinning structure that lies beneath the theoretical diversity manifested in rival and sometimes incompatible scientific accounts of the self and its aspects. I specify the basic structure of the self as embodied informational structures, that is, informational structures implemented in the mechanisms of information processing in cortical midline structures (CMS) and the Mirror Neuron System.

In the first chapter of the book, I survey four historical theories of the self and pin down the difference between substantivalism and its alternatives through some familiar philosophical examples. Two of these theories (by Aristotle and Descartes) exemplify substantivalism, whereas two others (by Hume and Kant) do not line up with substantivalism. I end the chapter by taking a fleeting look at significant breakthroughs that contribute to the formation of scientific psychology. In the second chapter, I survey SR in the philosophy of science, its origins, and its varieties. I survey the situation (in the history of modern physics) that motivated SR, and I argue that the theoretical diversity in the cognitive science, too, demands a structural realist strategy for addressing the problem of metaphysical underdetermination of the self. Chapter 3 provides more concrete examples to substantiate the last point about metaphysical underdetermination caused by the diversity of scientific accounts of the self. I mention theories of Metzinger, Gallagher, Georg Northoff, and Vittorio Gallese to show that there is indeed diversity in the field of theories of selfhood. In Chap. 4, I spell out SRS and specify the underpinning structure of the self in terms of embodied informational structures. I argue that it is possible to *represent* these structures mathematically through unifying Bayesian frameworks that are developed by Karl Friston in his statement of the free energy principle. I also argue that these structures are implemented in parts of the brain and organism (and according to an enactivist-ecological construal, the environment that situates the organism). Thus, although my philosophical account of the self is informed by scientific theories that evolved within a computational-informational framework, I explain that these structures are embodied in the brain and its world. In Chap. 5, I show how various accounts of consciousness, for example, the integrated information theory of consciousness, resting-state-based theory, and free-energy-based theory, could be unified by invoking a structural realist strategy, too. I also present a structural realist account of intentionality. Finally, in Chap. 6, I present a structural realist theory of the social aspects of the self, and I take the first steps to flesh out a structural realist account of moral aspects of the self. While I consolidate this account on the basis of experimental research on the Mirror Neuron System and Default Mode Network, I show that it is in line with some classical theories of morality, for example, by Peter Railton.

In all, in this book, I show that SRS provides a scientifically informed alternative to substantivalism. It endorses a structuralist account of basic, phenomenal, social, and moral aspects of the self. I argue that it is reasonable to take a modest realist attitude towards the self on the basis of what our best theories of cognitive science reveal about it, without succumbing to substantivalism.

SRS has evolved gradually through a number of papers that I have published previously. The feedback (and encouragement) that I received on those papers inspired me to work out connections between my different engagements with the philosophy of selfhood. I have aspired to pursue the ambitious project of presenting a comprehensive structural theory of selfhood which includes basic, phenomenal, and social aspects of the self. In this vein, I have to thank Shaun Gallagher for his earlier feedback on a couple of papers, and Tim Crane for his kind reception of my engagement with his theory of intentionality. The project began with a paper in Synthese, which provided structural realist construal of the CMS theory. Georg Northoff received my work very kindly and provided valuable feedback. I thank Georg sincerely for our fruitful dialogue about various subjects in cognitive science and philosophy. Also, I am indebted to Karl Friston, who patiently penned insightful comments on my earlier papers on predictive coding and the free energy principle. It would have not been possible to compose Chap. 5 without Friston's comments on an earlier draft that is incorporated into the chapter. All of these debts are gratefully acknowledged. Steven French and Steve Elliot replied to some technical questions about the publishing business, for which I am grateful (the name of the book is suggested by Steve Elliot). And I thank Lauriane Piette (who, in addition to usual editorial duties, kindly helped with copy-editing), the anonymous reviewer, and the rest of the Palgrave Macmillan team for their nice treatment of the project. I thank Dr.

Hossein Karami, the director of the Science and Technology Programme at the Amirkabir University for his support during my service there. As ever, I am hugely indebted to my wife Maaedeh and my parents for their support.

Nur-Sultan city   Majid D. Beni
Summer 2019

Praise for *Structuring the Self*

"Structural realism is now a well-established position in the philosophy of science. Majid Beni's rich and engaging work extends it from the physical sciences to neuroscience and from there to social relations and even our moral judgments. This is a bold and provocative move but Beni draws extensively on both recent and classic literature in philosophy and cognitive science to construct a compelling series of arguments. Structuralism has a long history spanning diverse fields of thought but here Beni breaks new ground and opens up further avenues of intellectual exploration that should be of interest to philosophers and scientists alike."

—Steven French, *Professor of Philosophy of Science, University of Leeds, UK*

"I have never read a book like this before: it offers a relentless and compelling tide of carefully constructed philosophical arguments about the very nature of 'self'. It bravely places philosophical arguments—well-honed over centuries—head-to-head with contemporary formulations of self-organisation in theoretical neurobiology and artificial (or perhaps artefactual) intelligence. The result is a convincing argument for a Structuralist Realist Theory of Selfhood—that neatly sidesteps the issues that attend eliminativism and substantivalism. This theory prescribes the right kind of ontological commitments to gracefully accommodate current formulations of consciousness in terms of information theory, neurobiology and the physics of sentient systems. As a physicist, I often find myself using phrases like 'self-organisation', 'self-assembly', 'self-information' and, latterly, 'self-evidencing'. After reading this book, I will never use the word 'self' in quite the same way."

—Karl J. Friston, *Scientific Director: Wellcome Centre for Human Neuroimaging, Institute of Neurology, UCL, UK*

"This is an impressive book about the self that fills a deep void in our conceptualations of ourselves. Bringing together, philosophy, neuroscience, and computational modelling, Beni develops a structural realist account of self that seems highly natural and plausible in metaphysical, conceptual, and empirical terms. A wonderful book that will set new directions and standards on the eternal debate of self."

—Georg Northoff, *EJLB-CIHR Michael Smith Chair in Neurosciences and Mental Health, University of Ottawa Institute of Mental health Research, Canada*

Contents

Abbreviations

AI	Artificial intelligence
CMS	Cortical midline structures
CSR	Cognitive structural realism
DMN	Default Mode Network
ESR	Epistemic structural realism
ISR	Informational Structural Realism
MNS	Mirror Neuron System
OSR	Ontic structural realism
SR	Structural realism

1

The Self, Its Substance, and Its Structure: A Selective History

1.1 Introduction

This chapter aims to flesh out the difference between the substantivalist, eliminativist, and structuralist approaches to understanding the self. It accomplishes its goal through surveying some deliberately selected fragments of the long history of philosophical engagements with the self-notion.[1] The chapter does not aim to review every important philosophical engagement with the self, much less to provide a comprehensive history of the philosophy of selfhood. There are various important theories of the selfhood—authored by Plato, Averroes, Leibniz, and John Locke, among many others—that are either glossed over or completely neglected in this chapter. Instead, the chapter includes a selective historical review which aims at fleshing out a significant philosophical difference between three ways of thinking about the self. These three approaches consist of the substantivalist account, the eliminativist account, and the structuralist account of the self. I embark on explaining these three ways of conceiving

---

[1] The notion of "self" has been used interchangeably with the notions of "mind" and "person" through the history of philosophy. In this book, too, I am usually using these three notions interchangeably (unless remarked otherwise).

© The Author(s) 2019

M. D. Beni, *Structuring the Self*, New Directions in Philosophy and Cognitive Science, https://doi.org/10.1007/978-3-030-31102-5_1

of selfhood through a quick survey of ideas of a few of the greatest philosophers of all ages. After briefly referring to some pre-Socratic speculations at the dawn of history, I will survey Aristotle's and Descartes' theories of selfhood. Despite their differences, I argue that the theories of Aristotle and Descartes somewhat agree in supporting a substantivalist conception of the self. I delineate the substantivalist view by referring to excerpts from works of Aristotle and Descartes. I also aim to scrutinise the solidity of grounds for adopting the substantivalist view throughout, and I introduce an alternative to substantivalism. As a foil to the substantivalist view, I highlight parts of theories of David Hume and Immanuel Kant. There is a difference between Hume's and Kant's respective views on the self; while Hume's theory may initially seem eliminativist, Kant defended a modified form of substantivalist theory of selfhood in the context of his transcendental approach. I argue that despite their difference, Hume's and Kant's respective theories could be understood as being at odds with the orthodox substantivalism of Aristotle and Descartes. Moreover, I argue that it is possible to construe both Hume's and Kant's conceptions of the self in the spirit of structuralism. Thus, I built upon examples from the history of philosophy to flesh out the distinction between the substantivalist, eliminativist, and structuralist perspectives. I draw on works of some of the most renowned philosophers of history to provide the reader with familiar examples of the orthodox foil (e.g., the Cartesian view) to structuralism as well as some familiar precedents (in the works of Hume and Kant) that are in line with my own theory of structural realism about the self.

Although I will allude to some exegetical points (in connection with works of Hume and Kant) to flesh out the structural realist theory of the self, my endeavour in this chapter is not completely exegetical. It is not my aim to show that structuralism is the only (or the most) plausible way of understanding the philosophical theses of Hume and Kant. Nor am I willing to suggest that Hume and Kant had been committed to structuralism as a philosophical agenda, in the same way that some contemporary philosophers, for example, John Worrall, Steven French, among others, are committed to it. The main reason for drawing on the theories of Hume and Kant in this chapter is that they both argue that the orthodox substantivalist theory of the self—as being developed by

Descartes—cannot be defended on the basis of either experiential or logical arguments. Despite such instances of scepticism about standard substantivalism, neither Hume nor Kant eliminates the self from their philosophy completely. The remaining option, which is not committed to either substantivalism or eliminativism, is structuralism. In this chapter, I argue that it is possible to recognise some early versions of structuralist theories of the self in works of Hume and Kant. Thus I introduce structural realism about the self by drawing on well-known historical examples, and thereby prepare the reader's mind for the structural realist theory that I will spell out in the book.

It is important to notice that, despite flirting with the history of philosophy in this chapter, this book does not substantiate the structural realist theory of the self on the basis of historical considerations. Rather, it aims to flesh out a structural realist theory of the self on the basis of recent theories of cognitive science and computational neuroscience. And today, our cognitive science is developed far beyond the scope of Hume's armchair psychology and Kant's transcendental, a priori psychology. In the final section of this chapter, I outline some important breakthroughs that contributed to the formation of cognitive science as a collaborative and multidisciplinary enterprise. In the next chapter, I will build the structural realist theory of the self upon recent findings in cognitive sciences.

1.2   In the Beginning

The famous Delphic proverb "Know thyself" has been attributed to a number of different sages of the days of yore. Even if the historical root of this proverb could be traced back to a particular person successfully, it would be still impossible to determine the precise historical moment in which a person, or more probably, a society of persons had begun to wonder about their identity. What is certain is that since time immemorable, "who am I" has been a tantalising question. To the extent that the official history of philosophy is concerned, some pre-Socratic philosophers were the first to voice such curiosity.

Even before the dawn of philosophy, there have been mythological theories of the selfhood. However, according to the surviving historical fragments, Thales of Miletus was the first person to enquire about such things. He enquired as to the nature of the world and the mind (or soul), by employing a philosophical (in contrast with mythological) methodology. According to Aristotle's report, Thales had said that "the loadstone has a soul because it moves the iron" (Graham, 2010, p. 35 A22). This assertion does not seem to be based on anything like scientific observation in the strict sense.[2] However, at least Aristotle construed Thales' phrase in a way that could make it intelligible to the contemporary reader. According to Aristotle, Thales had presumed that the soul is the source of motion (Graham, 2010, p. 35 A22). Aristotle's construal indicates that Thales has specified the notion of the mind on the basis of its observable function, namely being the cause of the motion. This means that Thales' theory was reliant on the pieces of evidence about the observable effects of the soul. He had endeavoured to support his stipulation of the soul as an unobservable entity, which is in possession of traceable causal powers by such pieces of evidence. The stipulation could explain the observable phenomenon—that some things are being moved—and there is a causal mechanism to explain the relationship between the soul and observable phenomena, given that the soul is the cause of motion. Thus Thales provides a functional analysis of the soul. It is true that he had also drawn a connection between the soul and water—which had been supposed to be the source of everything—by remarking that the mind made everything out of the water (Graham, 2010, p. 35 A23). However, as far as surviving fragments indicate, his attempt at characterising the soul as the cause of motion does not rely on mysterious speculation about the intrinsic nature of the soul. There had been others, aside from Thales, who had adopted this rather functional characterisation of the soul and assumed that the soul is the source of motion. As we will see in the next section, Aristotle himself advocates such a view. This is in contrast with Plato's various remarks (e.g., in *Meno, Phaedo*, and *Republic*) on the incorporeal nature

---

[2] This point would be even more conspicuous, if we remember that Thales had also asserted that the god is the mind or the soul of the world, the soul is mixed in everything, and all things are full of gods (Graham, 2010, p. 34 fragments A22–23). Despite its mystical air, as I explain immediately, Thales diverged from the mythological tradition, or at least so could it be argued, in the light of Aristotle's later development of Thales' approach (which could be recognised as somewhat functionalist).

of the soul which is supposed to explain the soul's capacity for understanding immaterial forms.

It is true that most of what pre-Socratics said about the soul remains enigmatic. For example, Thales asserted that the soul is mixed in everything, and all things are full of gods (Graham, 2010, p. 34 fragments A22–23). Some other pre-Socratics, such as Pythagoras, advocated more explicitly esoteric teachings about the soul, for example, its capacity for metamorphosis (or reincarnation) (Graham, 2010, p. 919). Some of the Platonic dialogues, for example, *Phaedrus*, preserved this air of mysticism. In comparison to this tradition, Aristotle provides a far more systematic and precise survey of the concept of the soul/mind/self. Aristotle's theory does not endorse mystical teachings such as reincarnation, nor is it openly committed to dualism—in the way that Plato's view on the incorporeal essence of the soul is committed. Despite its naturalistic tendency, Aristotle's view is philosophically too sophisticated and too loyal to the spirit of functionalism to reduce the soul to a natural element such as water or fire. Aristotle's theory introduces the soul in a naturalistic context and in relation to the natural life of organisms, but its thoroughgoing functionalism prevents it from identifying the soul with a material substance such as water. He does not explain the causal power of the soul in virtue of its material nature, for example, being a watery element. According to Aristotle, the soul is "a sort of first principle of animals" (Aristotle, 2016, pp. 1, 402a).

Aristotle's theory of the soul or self is important in the context of our present enterprise because it introduces a well-posed substantivalist theory of the self. According to Aristotle, the soul or self is a "substance", which bears certain properties or attributes and abides through changes and endures over time. I shall unpack this remark in the next section.

1.3   Aristotle's Substantivalism

For Aristotle, the world could be understood in terms of general categories or kinds, for example, quality, quantity, relation, place, date, posture, state, action, or passion (see Aristotle, 1963). This view holds that primary substances (e.g., a specific tree, a stone, a chair, etc.) are the basic units of existence, and primary substances have priority over other mentioned categories (quality, quantity, etc.). This means that the existence of instances of all other categories depends on the existence of instances of individual substances in which they inhere or to which they are related. Instances of each category must be ascribed to a specific substance or inhere in it as the properties, attributes, or accidents of that specific substance.

Aristotle's substantivalism has an intuitively appealing ring to it. Common sense indicates that the world consists of individual objects (or substances) such as trees, stones, chairs, fishes, and so on. And things must be distinguishable to be individuals. An individual object or a primary substance has some specific haecceity or thisness. The haecceity or thisness makes the object what it is, and it makes the object *essentially* discernible from all other things. A given tree has a thisness which makes it distinguishable from all other things in the world. Properties of the tree, its colour, age, size, and so on, exist in virtue of their relationship with (or inherence in) the tree. Aside from Aristotle's substantivalism, his hylomorphism, or his view on the relationship between form and matter, bears on his theory of the self. Below, I shall briefly elaborate on Aristotle's idea of hylomorphism.

Aristotle's theory of hylomorphism holds that a substance (i.e., an individual object) is a compound of matter and form. The matter is the stuff from which a thing is made. A tree is made of wood, a statue is made of bronze, and a candle is made of wax. The matter is a potentiality, whereas the form is actuality. Therefore, the matter does not confer on the substance a particular essence. It is the form of the tree that confers on it its haecceity, and it is the form of the statue which makes it an individual object, such as, say, the bust of Socrates. In unfolding his own theory of the soul in the second book of *De Anima*, Aristotle begins from a hylomorphic stance.

As Aristotle explains in his *De Anima*, it is possible to see some sort of substance as matter, but this sort of substance does not possess its thisness (is not a particular thing) in its own right. It is possible to understand a substance in terms of its essence or form, in virtue of which the substance has its thisness. And finally, it is possible to understand the substance as the compound of form and matter (Aristotle, 2016, pp. 22, 412a, 5–10).

After stating his stance, Aristotle explains that all natural bodies are substances in the third sense, as a compound of matter and form. Within this context, Aristotle submits that "the soul is a substance as the form of a natural body which has life in potentiality. But substance is actuality; hence, the soul will be an actuality of a body of such as sort" (Aristotle, 2016, pp. 22, 412a, 20). The organism becomes what it really is in virtue of its soul. This statement does not need, however, to lure us into the dualist puzzle of how to account for the relationship between the soul and the body. This is because the soul-body relationship is basically no different to the relationship between the form and the matter of a tree, or the sets of relations between forms and matters of all other things in the world. On the same subject, there may be grounds for defending a structuralist construal of Aristotle's theory. This is because Aristotle does understand "form" in terms of a thing's structure or principle of organisation. It could be pointed out that when he speaks of living things as having different kinds of souls, with different kinds of capacities, he emphasises that they have different kinds of souls because their matter is structured in different ways. In the same vein, it can be argued that when he speaks of the "formal cause",[3] he is referring to a thing's structure, that is, the way in which its parts are organised.[4] Indeed, he sometimes uses notions of "form" or "organisation" of an organism interchangeably with its "substance" (e.g., in *Metaphysics*, 1041a27–1044b22). If so, we can understand Aristotle's view on the self (or mind/soul) in terms of an interesting form of structuralism. But this is in contrast with Aristotle's emphasis on understanding substance as a compound of both form and matter (see my earlier remarks in the beginning of this section). I am generally sympathetic to the structuralist construal. But the relation between form and substance is not completely clear in Aristotle's works, and sometimes he suggests that primary substances as compounds of both form and matter have ontological priority (see my explanations earlier in this section). In view of Aristotle's emphasis on the status of the

---

[3] According to the Aristotelian view, in explaining a phenomenon, we have to specify four causes, that is, material, formal, efficient, and final causes. The body of an organism is enformed according to Aristotle, and the formal cause (the soul) explains the organism's capacity to grow (Aristotle, 2016, p. 411b).

[4] I thank the reviewer of this book for this remark.

primary substances as the building blocks of the world, I present his theory as substantivalism in this book. The ontic form of structural realism that will be outlined in the next chapter does not give ontological priority to individual substances.

Aristotle's account of hylomorphism may be generally wanting. One may question the grounds for making a distinction between the form and the matter, or one may feel that the relationship between the two is not explicated satisfactorily. However, Aristotle's account of the soul-body relationship is not particularly mysterious or problematic in comparison with other pairwise relations between forms and matters of other things. It does not dislodge the soul from the world, and the soul is related to the body in the same way that the form of the tree is connected with the material of the tree. On the same subject, it could be observed that by assuming that "the soul is not separable from the body" (Aristotle, 2016, pp. 24, 413a, 3),[5] Aristotle bypasses the platonic dualist tendency, avoiding the problem of how to connect the soul to the body. The soul and the body are inseparable because the soul is the essence and the first principle of the natural body. Neither bodies without souls nor souls without bodies are organisms. This does not mean that the soul and the body are the same thing. Aristotle considers the question of the identity of the soul and the body as unnecessary as the question of the identity of wax and its shape (say, in a candle, etc.). Suffice to say that, according to Aristotle, we can identify an ensouled being in virtue of its form or actuality, that is, its soul (ibid., pp. 412b, 7).

Let us recap. An organism as an actual living thing is a compound of the matter and form (or soul). The compound of the matter and form is a substance that endures through changes—a tree grows, a dog loses its teeth, a man becomes old—and bears certain properties—being green, being furry, being wise or foolish. Also, in virtue of their form or soul, living things have certain faculties. Different kinds of living things have different kinds of souls and thus different kinds of capacities. All living things have the capacity for nutrition, but plants have no other faculty besides nutritive faculty (see Aristotle, 2016, Book II, chapters 3–4). Since the soul is the source of faculties, even plants must have some kind of soul, that is, nutritive soul. This is because, as Aristotle argues, only an

---

[5] In addition to page numbers, I add the universal method of referring to *De Anima*.

ensouled body could possess the capacity for being nourished (ibid., pp. 31, 416a). Animals possess both nutritive faculty and perceptual faculty. Thus, in addition to having (potentially) the nutritive soul, an animal has a perceptual soul, in the sense that its organs are affected by the objects of perception. All animals have the faculty of nourishment—hence they have nutritive souls—and all animals have at least one kind of perception—hence they have a perceptual soul. The perceptual soul is the source of the will too, because all animals have the desiderative faculty, that is, appetite, desire, or wish (Aristotle, 2016, pp. 27, 414b, 5). In addition to having a nutritive soul and a perceptual soul, human beings also possess the reasoning faculty, that is, the rational soul, which is a part (or a kind) of the soul by which they know and understand things (Aristotle, 2016, Book III, chapter 4). For human beings, to comprehend things, their souls should be informed about things through perception. To be informed about things the part of the soul which is rational, that is, the rational soul, becomes "the place of forms", meaning that it catches the potential forms of objects (Aristotle, 2016, pp. 59, 429a, 29).

The relationship between the rational soul and the self must be analysed carefully. According to Aristotle, the soul is the form of the organism, and it bestows upon the organism its identity (thisness). Human beings owe their rationality to their rational souls, which bestow upon them their specific identities and make them particular substances. Thus, there should be a relationship between the rational soul of a human being and her/his self. It does not mean that the body is not part of the self though. According to Aristotle, an individual person, for example, Socrates, is a substance that is a compound of Socrates' body and the soul. That is to say, it is a compound of the body and rational soul actually and nutritive and perceptual souls potentially. I shall elaborate on the relationship between the soul and the self in the next section.

1.4   From the Soul to the Self

Aristotle advocates a substantivalist approach to what we call *the self* in this book. A human being, as a substance, is a compound of the body and the soul. According to his view, the self is the entirety of the body and the soul. Then again, Aristotle suggests that because the matter is a potentiality, it cannot contribute to the essence of selfhood, and it is the form that bestows upon a substance its identity or essence. In the same vein, it could be argued that because the person owes its thisness or haecceity to its form, it is possible to identify the self with only the soul and not the body. To be clear, it is possible to understand the self in an expansive sense, which includes the entirety of the body and the soul, given a hylomorphic distinction. However, it is also possible to understand the soul in a narrow sense, which includes only the soul as the essence of a person. Ronald Polansky has made the distinction between these two ways of understanding the self in his commentary on Aristotle's *De Anima*.

Polansky (2010, p. 4) argues that the expansive sense could be extended to greater things than the entirety of the soul and the body. The expansive sense of the self may be extended to include larger wholes, and it allows us to self-identify in terms of our community, or even the whole universe. Showing that the narrow understanding of the self is in line with a form of platonic dualism (in *Phaedo*), Polansky argues that "[f]rom this perspective the very self of a human being is the soul—the body being taken merely as something foreign to it—or some essence or Form, such as the Human Being itself, or even one's highest capacity, such as intellect" (Polansky, 2010, p. 3). To substantiate this point, Polansky draws attention to parts of *Metaphysics* and *Nicomachean Ethics* in which Aristotle identifies the self with the soul. However, given his commitment to hylomorphism, it should be assumed that for Aristotle the substance consists of both matter and form. Polansky argues that for Aristotle, the self is "the person or animal, composed of soul and body, rather than merely the soul that does things, such as experiencing emotions, perceiving, and thinking" (Polansky, 2010, p. 4). In this way, the self-substance is identified with the individual person which has a body and a mind (or soul), although it might be granted that the higher living beings, which possess rational soul, have more complex bodies, and the functions of their soul are somewhat less embodied (ibid., p. 5). Thus, the self is a substance which includes the entirety of the body and the soul. Here, I am not concerned with the correctness of either the expansive or the narrow construal of Aristotle's theory of the self. The important point is that both interpretations safeguard the substantivalist core of Aristotle's view of the self. For Aristotle, the self is an abiding and enduring substance.

In this book, I do not raise an issue with Aristotle's hylomorphism, which seems to be forsaken in the modern period. That is to say, even modern thinkers such as Descartes did not assume that the hylomorphism contributes to their philosophical discourse. Substantivalism, on the other hand, cast a long shadow on the later philosophical engagements with the notion of the self.

Aristotle's substantivalist view of the self might be intuitively appealing. It had appeared so for a long period, through the middle ages, and even in the modern period, where the essential part of the substantivalist view survived in the works of thinkers such as Descartes. Despite its intuitive appeal, Aristotle had not endeavoured to offer a specific argument for the plausibility of his substantivalist view of the self. He just stated the view and presented it as a natural way of conceiving of things that populate the world. He took the plausibility of the substantivalist view for granted. He assumed that primary substances, such as selves and other kinds of things, inhabit the world. The intuitive appeal of substantivalism was so strong that it remained in the very centre of the philosophical conception of the selfhood for a long time. Even thinkers such as Kant who pursued a revolutionary agenda sought to retain a modified form of substantivalism, which was restored in the contemporary period by thinkers such as P. F. Strawson. Aristotle did not argue for the plausibility of the substantivalist view, he took it for granted. Descartes, on the other hand, took great pains to substantiate his substantivalist view of the self on the basis of clear philosophical arguments. In the next section, I shall examine Descartes' endeavour for offering a specific argument for his substantivalist view of the self.

1.5   Descartes Ego, Existent or Non-Existent?

1.5.1   Je pense, donc Je suis

René Descartes has famously advocated a substantivalist account of the self. Cartesian philosophy parts ways with the Aristotelian theory on several occasions; Cartesian rationalism for example contrasts with the Aristotelian teaching according to which the intellect could not grasp things that have not been presented to it through senses (see Descartes, 2008a, p. part IV, 2008b, p. xx). Hylomorphism is another tenet of Aristotelian metaphysics that has been left out by Descartes. However, Descartes has retained Aristotelian substantivalism, albeit with some significant improvements. That is to say, Descartes remains committed to substantivalism, but he denies that physical substances are combinations of basic elements (air, fire, water, earth) or compounds of the form and matter—as Aristotle's hylomorphism indicates. Instead, he assumes that the world is made of homogenous stuff, and it possesses no quality beyond extension (Descartes, 2008b, pp. xix–xx). Despite differences with the Aristotelian view, Descartes still advocates a form of substantivalism. According to Cartesian substantivalism, there are two major kinds of substances (aside from God, as the primary substance). There is the material substance (*res extensa*) whose essence is extension, and then there is the mental substance (*res cogitans*) whose essence is cognition.

In his second meditation, Descartes examines various modes and qualities of a piece of wax to conclude that the essence of the wax is not its colour, shape, sonority, or any other transitive mode. According to Descartes, the piece of wax's essence consists in being extended (Descartes, 2008b, pp. 22–23). The same holds true for all other material things, whose essence is extension. All other properties or modes of the material substance, for example, its shape, motion, size, and so on, are accidental. Only the property of extension is essential to material substances. There are also immaterial substances whose essences are not extension. The essence of immaterial things, for example, Descartes' self, is thinking (Descartes, 2008b, p. 32). Thinking includes faculties of intellection, imagination, and will (Descartes, 2008b meditations four and six). Although Descartes presumes that the material substance and the immaterial substances are completely distinct, he argues that they are so closely conjoined that they form one single entity. Despite their close connection, body and mind are separate substances for Descartes. This is the thesis of Cartesian dualism, which presumes that the body and the mind are two distinct substances. The dualist has to provide a viable account for the relationship between the body and the mind.

There is an intimate connection between Descartes' argument for dualism and his insight into the existence of the self. Descartes' approach to demonstrating that the mind and the body are separate substances is introspective, in the sense that it is based on the assumption of first-hand cognitive access to the mental substance but not the physical substance. That is to say, Descartes relies on his intuition into his own mind (or self) and his mental faculties to conclude that the mental substance is distinct from the physical substance. According to him, "understanding, willing, doubting, and so forth are forms on the basis of which I recognize the substance known as *mind*; and I understand that the thinking substance is a complete thing just as clearly as I understand this of the extended substance" (Descartes, 2008b, p. 143). The important point here is that Descartes' belief in the existence of the mind as a separate substance is unshakable, and it lies at the centre of his philosophical system. Let me elaborate.

In his *Discourse on Method*, Descartes begins his speculation about the existence and duration of the mental substance (or the self) by completely rejecting anything that he can doubt, so as to find out if there remains anything that is completely indubitable. In this way, Descartes renounces all knowledge gained through sense perceptions, because sense perceptions could be erroneous. He also renounces knowledge produced by what he considers as valid proofs, because people make mistakes in reasoning and commit logical fallacies. This leads to the conclusion that the veracity of whatever had ever entered his head is as doubtful as illusions in a dream. There remains one unshakable assertion about the existence of the self. The truth of the assertion about the existence of the self cannot be questioned. As he notices,

> While I was trying to think of all things being false in this way, it was necessarily the case that I, who was thinking them, had to be something; and observing this truth: *I am thinking therefore I exist*, was so secure and certain that it could not be shaken by any of the most extravagant suppositions of the sceptics, I judged that I could accept it without scruple, as the first principle of the philosophy I was seeking. (Descartes, 2008a, p. 28)

The belief in the existence of the self-substance is indubitable. The self exists incontrovertibly, and introspection indicates that the essence of the self is thinking, intelligence, reason, or mind. The argument for dualism is based on the contrast between the indubitable existence of the self and the body on the one hand and the rest of the material world whose existence might be doubted on the other. According to Descartes:

> I was a *substance* whose whole *essence* or nature resides only in thinking, and which, in order to exist, has no need of place and is not dependent on any material thing. Accordingly this "I", that is to say, the Soul by which I am what I am, is entirely distinct from the body and is even easier to know than the body; and would not stop being everything it is, even if the body were not to exist. (Descartes, 2008a, p. 29)

The "I" or the "Soul" that is mentioned in this phrase might be recognised as a substantial self by Descartes. His argument for the distinction between the soul and the body could be challenged in a number of different ways. For example, his argument could be challenged on the grounds that one's ignorance of the sameness of the mind and body (or one's doubts about the existence of the body) does not warrant dualism. It may be argued that from his antecedents (his knowledge of the self, and his ignorance of its sameness with the body) it does not follow that mind and body are separate substances.

Be that as it may, I am not concerned with the dualism or its grounding. I am primarily concerned with the substantivalist conception of the self—the thesis that the self is a substance. That being the case, I do not aim to criticise dualism or address the mind-body problem (directly) so much as I endeavour to present a viable alternative to the substantivalist theory of the self. Substantivalism indicates that the self is an enduring individual substance. Aristotle advocated a substantivalist theory of the self, but he did not bother to substantiate his view with compelling arguments. Descartes, on the other hand, offered arguments for his substantivalist view. This is an important point in the history of the philosophy of the self, and we may begin our quest for finding a viable alternative to the substantivalist view by scrutinising Descartes' reasoning. I shall scrutinise Descartes' arguments in the next section.

1.5.2   Do *I* Exist?

The Cartesian scholarship includes well-known criticisms of Descartes' argument. For example, Bernard Williams challenges Descartes' argument by arguing that introspection cannot warrant ontological claims about facts of the matter. This means that the subjective report "I am thinking" does not entail an ontological claim about one's existence, in the sense of a real entity that inhabits the real world. Williams argues that "Descartes could not make an inference from 'I am thinking' to 'I exist', if that, in its turn, were taken to represent (as Descartes takes it to represent) a state of affairs which could be third-personally represented as '*A* exists' ('there is such a thing as *A*')" (Williams, 1978, p. 82). I can appreciate the insight behind objections that target the introspective core of Descartes' argumentation. However, my main reservation concerns the failure of Descartes' argumentation to substantiate the existence of a substance that abides over time. I shall unpack this remark immediately with reference to Gassendi's criticism.

In his *Disquisitio Metaphysica seu Dubitationes & Instantiae adversus Renati Cartesii Metaphysicam & Reponsa* (1644), Pierre Gassendi challenged Descartes' argument—that is, "I am thinking" therefore "I exist"—by raising the issue of the substance's endurance over time. The general insight behind Gassendi's criticism is that one must account for one's enduring identity over time in order to demonstrate that one is a substance. As I showed in the previous section, Descartes offers an interesting (though as I will argue, questionable) argument for the existence of himself, by reasoning that thinking and even doubting presupposes the existence of the self. However, he does not provide a well-posed account of the identity of the self over time. As my reference to Descartes in the previous section indicates, Descartes draws a connection between thinking (and even doubting) on the one hand and his being on the other. I can doubt the existence of everything, doubting is a form of thinking, and the fact that I can doubt shows that I can think. I should exist even if I want to doubt my own existence or to have any thoughts of any kinds. Because I am thinking, I exist. There are reservations about the sufficiency of introspection for establishing an existential conclusion.

However, even if we grant that "I am thinking" or "I have thoughts" can validly entail an existential conclusion, we cannot assert that changing thoughts are modes of the same enduring substance and that they can be ascribed to an abiding individual self. I shall unpack this point presently. Descartes' account of the relationship between modes (or properties) and the substance is in line with the good old substantivalist intuition according to which properties depend on their substance and inhere in it (Shapiro, 2012, p. 229). The substantivalist view also presumes that the substance endures through changes and abides over time. Accordingly, Descartes should assume that thoughts depend on the self-same individual substance that persists during various intellectual episodes (ibid., p. 230). However, Descartes does not offer arguments to substantiate his view on the endurance of the substance and its identity over time. This means that although Descartes argues for his substantivalist conception of the self—that is, by saying that "I am thinking, therefore I exist"—he does not develop his argument to demonstrate that what exists is an enduring and abiding self-substance.

According to Gassendi (op. cit.), from the fact one is thinking, it cannot be concluded that one is a particular thinker—in the sense of an individual substance that endures over time. To establish the point that various thinking episodes belong to an abiding self-same substance, Descartes must presume the existence of what he is supposed to prove, "namely, that there exists a particular person endowed with the capacity for thought" (Fisher, 2005, p. 14). Thus, Gassendi argues that the existence of cognitive activity, for example, thinking, doubting, desiring, and so on, is not enough for demonstrating that there exist such things as selves, in the substantivalist sense, that is, in the sense of relatively independent individual things that endure through changes. In short, Descartes' argument for the existence of his self as "I" or the "Soul" is not valid as it begs the question of the existence of the self over time instead of proving it.

It is possible to make a reconstruction of Descartes' argument invulnerable to Gassendi's objection. For example, it is possible to account for the endurance of the self over time on the basis of Descartes' remarks on the psychological dimensions of the self or its psychological capacity for remembrance. In this vein, Shapiro suggests that "the meditator's memory involves a kind of appropriation of his prior thoughts, and that these acts of memory serve to constitute the meditator as the same thing over time" (Shapiro, 2012, p. 236). However, this suggestion is based on a developmental conception of the Cartesian self, where the self is conceived as something that develops over time and owes its evolving constitution to the psychological faculty of memory. This construal seems to be inconsistent with the substantivalist view of the self, according to which the thinking substance is an invariable entity, which endures through the changes. According to the substantivalist view, changes do not affect the unvarying constitution or essence of the substance, and the substance remains the bearer of the properties and changes.

Let me recap. Descartes has endeavoured to argue for his substantivalist conception of the self. However, Descartes' argument for the existence of the self as a substance, that is, a basic and relatively independent individual object, that endures over time and bears various properties or modes has not been compelling. In the next section, I present David Hume's conception of the self. Hume theory of the self provides a viable alternative to the substantivalist views of both Aristotle and Descartes.

1.6   Hume's Scientific Philosophy, Eliminativist or Structuralist?

David Hume ([1738] 2000) parted ways with the substantivalist tradition. Both Aristotle and Descartes had regarded physical objects and persons as substances (though their notions of substance are not completely the same). Hume did not conceive of physical things and persons in terms of substances. His attempt at finding new foundations for philosophy led him to a theory that is essentially different from the orthodox substantivalist metaphysics. Hume's ([1738] 2000, p. 47) own project can be understood as a form of scientifically informed philosophy that was influenced by "Newtonian" breakthroughs of the seventeenth century. His project aimed to construct a complete system of the sciences, "built on a foundation almost entirely new, and the only one upon which they [i.e., the sciences] can stand with any security" (Hume, 2000, p. 4).

This foundation is laid down by what Hume calls "the science of man". The science of man unveils "the only solid foundation for the other sciences" (Hume, 2000, p. 4). Accordingly, "the only solid foundation we can give to this science itself must be laid on experience and observation" (Hume, 2000, p. 4). Since the rise of Newtonian physics in the seventeenth century, it has been usually supposed that physics sets forth the exemplar of a successful scientific research program which includes the foundation of all other sciences (Oppenheim & Putnam, 1958). However, I agree with the Humean insight according to which our scientific theories of the brain and cognition provide a secure foundation for structuring and unifying other pieces of knowledge (Beni, 2019). My project in this book is in line with this insight too. I provide a scientifically informed metaphysical alternative to orthodox substantivalism about the self. The alternative metaphysical point of view is informed by recent breakthroughs in computational neuroscience and cognitive sciences. Below I shall survey Hume's divergence from substantivalism about the self.

Being influenced by the success of Newtonian physics, which unfolds the universal law of gravitation, Hume acknowledges that sciences must engage in the discovery of universal laws and principles. Moreover, he suggests that the exploration of laws and principles of Human understanding is a valuable enterprise, which can underlay all other subdisciplines of knowledge. In this context, Hume embarks on the exploration of universal principles that guide the faculty of imagination. It is through the faculty's associative principles (viz. resemblance, contiguity in time and space and cause and effect) that associations between ideas arise (Hume, 2000, p. 13). The principle of association of ideas is at the centre of Hume's science of man—in the same way that the law of gravity is at the centre of Newton's physics. In his *Treatise*, Hume often appeals to this principle to account for cognition. According to him,

> 'Tis impossible for the mind to fix itself steadily upon one idea for any considerable time; nor can it by its utmost efforts ever arrive at such a constancy. But however changeable our thoughts may be, they are not entirely without rule and method in their changes. The rule, by which they proceed, is to pass from one object to what is resembling, contiguous to, or produc'd by it. When one idea is present to the imagination, any other,

united by these relations, naturally follows it, and enters with more facility by means of that introduction. (Hume, 2000, p. 186)

In the next two subsections, I explain how Hume explicates the concepts of physical and mental objects based on the principle of association of ideas, without invoking the notions of the physical and mental substances.

1.6.1   Eliminating Physical and Mental Substances?

Hume presumes that perceptions consist of impressions and ideas, with the difference that impressions are more forceful, lively, and vivid, and all ideas are ultimately copies of impressions. The mind can use perceptions to represent external objects. However, senses do not present their impressions as images of something independent, external, and distinct. Perceptions are what the mind perceives, and we can be certain only of the existence of perceptions, which are "immediately present to us by consciousness, command our strongest assent, and are the first foundation of all our conclusions" (Hume, 2000, p. 140). Although we can be certain of the existence of perceptions, perceptions themselves cannot reassure us that they are representing external objects. Therefore, the assumption of the continuous existence of physical object—called "double existence" by Hume—must be a result of the working of imagination or reason (Hume, 2000, p. 126). The assumption of the existence of *external* objects cannot be supported by reason, because reason cannot validly draw inferences from experience to what lies beyond the domain of experience. According to Hume, "we may observe a conjunction or a relation of cause and effect betwixt different perceptions, but can never observe it betwixt perceptions and objects. 'Tis impossible, therefore, that from the existence or any of the qualities of the former, we can ever form any conclusion concerning the existence of the latter, or ever satisfy our reason in this particular" (Hume, 2000, p. 141). For the same reason, the assumption of the continuous existence of physical objects cannot be corroborated validly by imagination. This is because while imagination and its principles, for example, of cause and effect, are always reliable in setting relations between perceptions, they cannot reliably draw connections between perceptions and unperceived continuous objects. In this fashion, Hume dissolves the assumption of the external existence of physical substance. He submits that for him "there is only a single existence, which I shall call indifferently *object* or *perception*, according as it shall seem best to suit my purpose, understanding by both of them what any common man means by a hat, or shoe, or stone, or any other impression, convey'd to him by his senses"(Hume, 2000, p. 134). In brief, Hume dispenses with the notion of physical substance. As I shall indicate, he also dissolves the notion of self-substance.

The substantivalist view of the self presumes that the self is an individual entity that endures through changes and is the bearer of properties or modes, for example, its perceptions, thoughts, desires, and so on. Against this substantivalist view, Hume argues that thoughts and perceptions may exist separately and independently, namely without being related to a substance. He denies that perceptions as properties need to be borne or supported by an underlying substance. His argument against the assumption of the existence of self-substance is essentially the same as that which grounds his scepticism about the existence of physical substances. The general insight behind this naysaying is that we have no evidence (i.e., no impression or perception) for the assumption of the existence of self-substance. And without the requisite pieces of evidence (based on impressions), imagination and reason cannot ensure the existence of the self-substance. It follows that in the absence of compelling evidence, perceptions do not need to be related to or borne by a substantial self. As Hume argues:

> For my part, when I enter most intimately into what I call *myself*, I always stumble on some particular perception or other, of heat or cold, light or shade, love or hatred, pain or pleasure. I never can catch *myself* at any time without a perception, and never can observe any thing but the perception. (Hume, 2000, p. 165)

In this fashion, Hume replaces the substantivalist theory of the self with a bundle theory, which identifies the self with a bundle of perceptions. There is no single impression that corresponds to our idea of the self-substance. Thus, Hume completely denies that the "substance/property" distinction can be included in a viable vocabulary for speaking about the self. He argues that "the theory of substance merely crowds the mental realm with something totally inaccessible, whose relation to perceptions ('inherence') remains utterly mysterious" (Brett, 1990, p. 24). Does this mean that Hume eliminates the self from his philosophy altogether? I shall address this question in the next section.

1.6.2   An Early Version of Structuralism About the Self

Hume cannot deny the existence of the self without being involved in a paradox. As Brett argues:

> If we take Hume to be denying either the existence or the identity of the self or mind when he talks about fictions and mistakes, then the account seems degenerate. We end up with nothing which can function as the possessor of this fiction and nothing to which we can ascribe the activities or dispositions which generate this error. (Brett, 1990, p. 26)

The reservation behind Brett's judgement is understandable. If we assume that there are no such things as selves, we will not find anything on which to hang the allegation of being erroneous about the existence of the self. The assumption of the existence of the self could be erroneous, and the self could be nothing but a fictional entity. But short of introducing a viable substitute for the self, we will be trapped in paradox when we suggest that the assumption of the existence of the self is erroneous because we cannot implicate who has committed the error. If we assert that the self-concept is a fiction, we cannot say who told the fiction, and we cannot implicate the entity that committed the error of presuming that the fictional entity is real. So, if we affirm that the self does not exist, we cannot assert that the self-concept is fictional and judgement about its reality is erroneous. Hence the paradox.

To recap: on the one hand, Hume argues that the assumption of the existence of self-substance cannot be plausibly defended. On the other hand, he cannot eliminate the self-substance without being trapped in a paradoxical situation (let us leave out the physical substance for the time being). From the structural realist perspective that I will defend in this book, the natural solution is to dispose of the substantivalist conception of the self, but retain a modified (non-substantial) version of the self. In the next chapters of this book, I will embark on specifying the non-substantial conception of the self along the lines of structural realism. In the reminder of this subsection, I will argue that we may reasonably attribute a structuralist conception of the self to Hume.

Hume denies that we can find impressions that correspond to physical and mental substances, and thereby eliminates physical and mental substances from his philosophical system. I am not concerned with the existence of physical bodies in this book, so let us focus on selfhood. Does the elimination of the self-substance indicate that Hume denies that there are such things as selves? As I remarked in the previous section, he retains a form of a bundle theory of selfhood, according to which the self consists in a bundle of perceptions, for example, heat or cold, pain or pleasure, and so on. The question is, what is the thread which sews the patches of the bundle together? It is true that the bundle theory does not presume the existence of the self-substance. However, even the bundle theory presumes that perceptions are not completely separate and independent from one another. Perceptions must be connected to each other to form a bundle after all. Thus, even if we want to advocate a bundle theory of the self, we have to find a way to draw meaningful connections between at least some series of perceptions (Brett, 1990, p. 24). For, a bundle would not exist unless its patches could be put together. The situation tends to get paradoxical. On the one hand, Hume presumes that perceptions are distinct and identify with independent existences, in the sense that they are not dependent on a mental or physical substance. On the other hand, even the bundle theory should grant that perceptions "form a whole only by being connected together" (Hume, 2000, p. 400). The question is how could diverse and separate perceptions contribute to forming a bundle of interrelated perceptions without being related to or borne by a substantivalist self? I argue that Humean philosophy allows for specifying the real connections between diverse perceptions in terms of the integrating force of the association of ideas. This solution would allow for integrating various perceptions together without invoking the self-substance. I argue that it is possible to construe this approach in terms of structuralism.

Obviously, if we remain committed to the perspective of the orthodox (substantivalist) metaphysics, we cannot account for the real connection between perceptions without relating them to a substance. From the substantivalist point of view, the modes of the self-substance, for example, perceptions, could not contribute to forming a self without being underpinned by a substratum. However, as I will argue in the next chapter, contemporary philosophy of science includes a viable structuralist alternative to orthodox substantivalist metaphysics. From this new structuralist perspective, objects do not need to be necessarily categorised in terms of either a substance or something borne by a substance (e.g., a property). This view allows us to conceive of things, for example, the self, in terms of structured units instead of the orthodox substances. Below, I briefly explain how it is that Hume's approach allows us to construe the self in terms of a structured unit.

At times, Hume's philosophy has been unfavourably evaluated as an attempt at spreading scepticism. However, as my brief elaboration at the beginning of Sect. 1.5 indicates, Hume aims to launch a scientific-minded enterprise for unveiling the most fundamental principles of human understanding. Accordingly, Hume endeavours to account for the existence of things such as physical objects or selves without invoking substances and quiddities, whose essence would remain mysterious. Instead, he endeavours to construct his account of knowledge (of things and selves) on the fundamental principles of understanding, which are unveiled by his science of man. Accordingly, he explicates the conception of the self in terms of principles of association of ideas, and especially the faculty of causation, which can forge self-patterns without demanding the existence of anything like a traditional substance. It follows that it is possible to assume that there are such things as selves (even without making commitments to substantivalism), provided that we concede identifying the selves in terms of the strength of connections between perceptions. The force of association of ideas forges the patterns of real connections that relate distinct perceptions. This construal identifies the unified self in terms of a structure of perceptions related together via patterns of causal connections. The self is nothing but the dynamical structure that confers on the constellation of patterns of perceptions its various configurations over time. Please note that, despite the prima facie plausibility of this reply, Hume did not develop a full-blown philosophical alternative to the orthodox metaphysical account of the self. Under the circumstances, it could still be contended that dynamical structure itself should be borne by or related to a substantial self. However, the ontic structural realism, which will be outlined in the next chapter, denies that the individual substances have ontic priority over structures. According to this view, the assumption of the ontological priority of substantial self is not inevitable. To the extent that experimental evidence is at issue, this assumption (about the priority of individuals) is unjustified. The alternative structuralist metaphysics that will be unfolded in the next chapter gives ontological priority to the self-structure and submits that the existence of the dynamical structure *does not need to be supported by individual substances*. The structuralist view holds that perceptions are what they are in virtue of featuring in the patterns of experiences in a particular way, and the structure of the self could be identified as the basic structure that underlies patterns of causal relations that connect various perceptions.

According to my structuralist reading of Hume's theory, the self may well exist, but not as an orthodox substance. It may exist as the structure that underlies patterns of perceptions, which are forged through the exertion of the power of association of ideas and especially through the associative principle of cause and effect. The role of causation is important because according to Hume, causation plays an important part in transmitting the existence through the ontological web. Hume submits that: "[t]he only conclusion we can draw from the existence of one thing to that of another, is by means of the relation of cause and effect, which shows that there is a connexion betwixt them, and that the existence of one is dependent on that of the other" (Hume, 2000, p. 140). In this fashion, Hume can explicate the self-notion in terms of the causal patterns that subsume real causal connections that unify distinct perceptions and make them self-related. Through the web of causal connections, self-related perceptions can be related to one another and constitute the structural self, without being dependent on the self-substance or a subject of inherence. Because this construal identifies the self in terms of the patterns of causal connections between distinct perceptions, it can be understood as a structuralist theory of the self. I think my structuralist reading is in line with some existing interpretations of Hume's theory of the self. Nathan Brett's (1990) causal theory of the self is such an interpretation. Brett argues that the self or the mind "is a set of perceptions which are causally integrated and whose causal integration is specified by the principles of association. Thus, there is a sense in which a mind is, not just a succession of perceptions; it is a causal association of perceptions" (Brett, 1990, p. 26). This interpretation adds up to accounting for the unified self as a causally integrated bundle of perceptions, without invoking the orthodox self-substance.

Let us recap. Real connections between diverse perceptions (i.e., diverse existences) could be specified in terms of causal patterns. It might still be objected that while the perceptions are real distinct existences, causal links are not real enough to provide real connections between distinct existences. However, it should be noticed that this understanding has its roots in the orthodox substantivalist view, according to which only the individual objects could be identified with constituents of reality. A structuralist does not have such scruples. A structuralist theory of the self allows us to understand causal connections as real constituents of what there is. This is in line with Brett's construal which holds that "being real" is not in contrast with "being a relation". Perceptions are distinct existences. However, there is no reason to assume that causal relations are not real connections (Brett, 1990, p. 30). The self is the underlying structure that subsumes causal patterns of perceptions. It imposes order on distinct mental entities, for example, perceptions. There remain important exegetical questions that have to be addressed before we could establish the plausibility of the structuralist construal of Hume's theory of the self. However, the point of this chapter is not exegetical. I refer to Hume's work to shed light on my own stance on the self. When fleshing out my view in the next chapter, I shall go beyond exegetical and historical cases and draw on recent scientific breakthroughs to support my structuralist account. Before that, I also outline Kant's sophisticated engagement with the substantivalist theory.

1.7   The Kantian Account of the Self, Substantivalist or Structuralist?

The concept of substance is to some extent exonerated in Kant's philosophy. This means that Kant retains a thin and modified notion of the self-substance by suggesting that the unity of the subject of experiences is the transcendental condition of having experiences at all. However, because Kant denies that (either the physical or the mental) substances are objects-in-themselves, his account of the self diverges from the orthodox version of substantivalism severely. Kant grants that substances are constitutive, but only in the sense that they are necessary conditions (or a priori categories) which make our experience of objects possible. Our cognitive constitution enables us to understand physical objects as units of existence that endure through changes and bear various properties. This substantivalist conception of physical objects is formed through analogy with our ways of thinking of ourselves as substances that endure through changes and bear properties and thoughts. Alternatively, it could be assumed that we conceive of ourselves as substances in analogy with the way in which we conceive of physical objects (Kant, 1998, p. 54 Guyer and Wood's introduction). Be that as it may, the concept of self-substance is *transcendental* in the sense that it does not come from experience; it provides the a priori and necessary condition of experience, where experience is the result of dynamical interactions between external objects and our cognitive faculties. We have to appeal to the self-concept to explain why we experience the world and ourselves in a normal way.

Kant does not presume that we could know things-in-themselves. However, he does not preclude the possibility of knowing facts *about* things-in-themselves. Although we cannot directly access the world as the totality of things-in-themselves, Kant claims that we can discover the structure of our own conceptual scheme and the categories that govern our experience of the world. Conceptual categories present the necessary conditions of our experiences. According to Kant (1998, p. 212), there are four classes of categories, that is, quantity, quality, relation, and modality. Each class consists of three pure concepts of understanding. Pure concepts of inherence and subsistence are classified under the category of relations. Pure concepts of inherence and subsistence provide the notions of substances and accidents. In short, the notion of substance is a pure concept of understanding, which makes our experience of objects (i.e., physical objects and selves) as unified particulars possible. This pure concept is merely a condition of understanding, and it has no objective validity as a content of experience or cognition. That is to say, we cannot make meaningful assertions concerning substances as things-in-themselves. I understand the self (myself) as a substance not because the self-substance is a real object. The Kant-specific argument for this claim is complicated and at times confusing. Below, I shall endeavour to canvass it.

We can experience objects as unified wholes. Kant submits that the experience of the unity of experiences demands the unity of consciousness (Kant, 1998, p. 232). For the diversity of the data of intuition to be unified (and thereby perceivable), the subject's consciousness must be unified. This condition—that is, the unity of consciousness and thus the unity of the self—precedes all experiences. It is an a priori, transcendental condition. Kant specifies this condition in terms of the transcendental apperception, which confers on the consciousness its unity and thereby grounds all experiences (Kant, 1998, pp. 232–233). Kant is clear that empirical self-consciousness, as grasped by inner sense or introspection, cannot ground an abiding notion of self-substance; we cannot "intuit" the abiding subject of experience or the self-substance. Kant grants that we can be *conscious* of the self as the subject of our experiences, but this does not mean that we could have any knowledge of the real nature of the enduring and abiding self-substance. It may seem confusing that Kant grants the possibility of being conscious of the self but denies the possibility of its cognition. However, being conscious of the self, say, through the inner sense, is not enough for the knowledge of the self-substance as the content of experience. The condition of the unity of consciousness is transcendental, and it does not have a knowable content. As Kant argues, "[t]he consciousness of oneself is therefore far from being a cognition of oneself, regardless of all the categories that constitute the thinking of an object in general through combination of the manifold in an apperception" (Kant, 1998, p. 260). Kant's argumentation is complicated.

Nevertheless, it explicitly indicates that I cannot know myself as I (as a substance) really am, but only as I appear to myself (ibid.).

To wit, Kant makes room for the self-substance, because he still presumes that the notion of the substance is a constitutive component of the experience. However, his notion of the self-substance is thin. The substance is nothing but a transcendental condition of experience. From Kant's conception of substance, it does not follow that there are such things as selves as real things-in-themselves. In this sense, Kant's substantivalism is essentially different to both Aristotelian and Cartesian forms of substantivalism discussed in previous sections. Kant presumes that experience or empirical self-consciousness does not warrant our belief in the existence of the substantial self. Nor is there a compelling rational argument for the existence of self-substance. Kant's criticism of the paralogism of pure psychology in (the first edition of) his *Critique of Pure Reason* is presented to this effect, namely to show why the existence of self-substance cannot be even deduced rationally. Below, I shall cite Kant's criticism.

The classical (Cartesian) argument for the existence of the self-substance was presented in the following way:

1. The substance is the absolute concept of our judgments
2. I, as a thinking being, am the absolute subject of all my possible judgments
3. It follows that I, as a thinking thing, am a substance (Kant, 1998, pp. 415–416).

According to Kant, this argument is invalid. It is true that I can think of myself as a substance, in the sense that I can distinguish myself from mere predicates and determinations of things. Indeed, given that the substance is a pure concept of understanding, I cannot help conceiving of myself as a substance. However, from this observation, it does not follow that I, as a thinking thing, am a real object that endures through changes and abides over time. This means that the paralogism of pure psychology fails to prove that the self-substance endures through all alterations. Please note that the paralogism of pure psychology presumes that only the notion of self-substance could guarantee the endurance and identity of the substance over time. Kant somehow acknowledges this point and agrees that the only useful point in ascribing substantiality to the self it that it may help us to account for the self's identity over time. According to Kant, if we cannot substantiate this point, we "could very well dispense with it [the substantivalist notion of the self] altogether" (Kant, 1998, p. 416). And because, as Kant argues, the orthodox version of substantivalism cannot account for the endurance of the substance over time, we can as well dispense with the self-substance. Kant's own alternative version of substantivalist is remarkably modest. From the mentioned criticism, he concludes that:

> From this it follows that the first syllogism of transcendental psychology imposes on us an only allegedly new insight when it passes off the constant logical subject of thinking as the cognition of a real subject of inherence, with which we do not and cannot have the least acquaintance, because consciousness is the one single thing that makes all representations into thoughts, and in which, therefore, as in the transcendental subject, our perceptions must be encountered; and apart from this logical significance of the I, we have no acquaintance with the subject in itself that grounds this I as a substratum, just as it grounds all thoughts. (Kant, 1998, p. 417)

As the quotation is indicating, Kant defends a form of philosophical humility according to which we do not have any knowledge of the intrinsic features of the self as a substance or its real essence. So, although Kant retains a thin notion of self-substance, as a form of a pure concept of understanding, he does not endorse the presumptuousness of the orthodox substantivalism, according to which we could know the real subject of mental inherence.

There are of course various ways of making sense of Kant's noncommittal approach to the existence of the self-substance as an individual object-in-itself. For example, Arthur Melnick has argued that it is best to understand the Kantian self not as an entity but as an intellectual or intelligent action that emerges as an "engagement or marshalling of the intellectual faculty" (Melnick, 2010, p. 4). According to this construal, thinking is inchoate, unsettled, and unformed before being marshalled and coalesced around a series of particular thoughts (ibid.). This construal, too, indicates that the self is not the substance (in either the Aristotelian sense or the Cartesian one). It is the intellectual marshalling, rather than the self-substance, which structures the thoughts. Although I am sympathetic with Melnick's construal, I do not defend his view on the nature of the self as intellectual marshalling coalescing. I agree, however, that it is best not to see the self as an individual object or substance but as a dynamical structure that changes its configuration over time.[6] I mention his construal nevertheless because this reading points to the self as a dynamical structure, which is at the centre of our enterprise in this book. Melnick's interpretation specifies "transcendental self-consciousness" in terms of "a dynamical structuring within intellectual marshalling in coming to make the report" (Melnick, 2010, p. 54). The dynamical structuring does not demand the existence of the orthodox self-substance, and yet it accounts for the unity of our experiences of the multitude of data of intuition.

It is true that at times, it has been assumed that the Kantian approach does not aim to dispense with orthodox substantivalism so much as to make an improvement upon it. A version of such a substantivalist construal of Kant's theory of objects and persons has been reconstructed by P. F. Strawson. In his *Individuals*, Strawson (1959) has developed a Kantian version of what he calls descriptive metaphysics. According to this view, in order to understand the world, we have to acknowledge that the world consists of physical substances and individual persons (as self-substances). So, at least some interpretations of Kant construe his theory of the self in terms close to what the orthodox substantivalism offers. But the substantivalist construal of Kant's theory of self is not in harmony with the structural realist approach that I develop in this book (in Chap. 3, I explain my reasons for disagreement with Strawson's substantivalist

---

[6] To be clear, I am particularly sympathetic to parts of Melnick's construal that specify the self as a dynamical structure. For example, take the following phrase:

> There is always a structure of clustering or coalescing of inchoate thoughts about any thought in "peak" condition of being fully formed as a specific comprehension). There may be one or several such clusters depending on whether my marshaling is in regard to having a single thought or several different thoughts. It is such a structure to the marshaling that gives a distinction between the thoughts and the subject that is settled on them, holding them, etcetera. Although such clusters can "divide" and enter into subsequent marshalings, they can only do so by altering their degree or intensity (how strong the settling is, how intense the hold is). (Melnick, 2010, p. 48)

construal). Instead, I defer to the structuralist construal of Kant's theory of the selfhood according to which "Kant's thinking self as a form or structure that eludes any attempt at reification" (Zoeller, 1993, p. 460). Notice that, although Kant's critical philosophy excludes the possibility of reification of the self (or the cognition of its true nature), it does not preclude the possibility of knowing facts about self, especially (according to my construal) facts about the basic structure of the self. This point is worth underlining, because as I will explain in the next chapter, there are some forms of contemporary scientific structuralism (in philosophy of science) that mesh well with this Kantian teaching, according to which we can only know the structural features of reality, but not the intrinsic features of objects or things-in-themselves. In this book, I will project this structural realist approach to the philosophical conception of the self. The structural realist theory of self dispenses with the orthodox substantivalism but does not eliminate the self from ontology and epistemology completely. This theory of the self is in line with a structuralist understanding of Kant's view of the self. On the one hand, he denies that the self-substance could be identified in terms of an abiding and enduring subject of experience, in the sense that is at issue in the orthodox substantivalism. On the other hand, Kant uses a transcendental argument to retain a thin notion of self-substance in his transcendental psychology. Despite general agreement with Kant's view, my main strategy for defending a structural realist theory of the self is not strictly Kantian. Unlike Kant, I do not conjure any transcendental arguments to flesh out my views. I mainly endeavour to present a philosophical theory of the self on the basis of what our best theories of contemporary cognitive sciences tell us about the self.

1.8   Some Recent Scientific Breakthroughs, an Overview

As I have remarked in Sect. 1.5 of this chapter, Hume regarded his project in *Treatise* as an endeavour for developing what he called the "science of man", a science which was allegedly based on observation and experimentation. This science was supposed to provide some foundations for other sciences. Using the word "science" became fashionable in the context of scientific breakthroughs of the seventeenth century—for example, Kepler's discovery of laws of planetary motion, Harvey's discovery of blood circulation, Galileo's laws of falling body, Boyle's laws of ideal gas, Newton's description of the universal gravitation, among other advancements. Hume's attempt at finding fundamental and unifying principles of his science of man aimed to match Newton's endeavour which had led to the discovery of laws of motion. Hume's scientific-mindedness is quite obvious throughout his *Treatise*. Aristotle's, Descartes', and Kant's respective theories of mind and cognition had been evolved as pieces of methodological reasoning supported by observations that had been available to those thinkers at the time. However, none of these philosophical enterprises (not even Hume's scientific project) was scientific in the strict sense, say, in the sense that Newtonian physics was scientific. Psychology has progressed a lot since the time of Hume and Kant. In the nineteenth century, some notable neurophysiologists and psychologists, for example, Herman von Helmholtz ([1910] 1962), Wundt (1980), and James (1890), made great strides towards a scientific discipline of psychology. Despite some affinities with the works of philosophers such as Hume and Kant, the emerging empirical psychology was rather different from both Hume's armchair science of man and Kant's transcendental approach to cognition. This empirical approach to psychology has thrived in the twentieth century.

My philosophical theory of the self in this book is informed by recent breakthroughs in cognitive science—as a multidisciplinary enterprise with links to psychology, neuroscience, computer science (including artificial intelligence or AI), and various other disciplines. There is no space here for a detailed review of the history of the development of cognitive science. Such a review has been offered by Bechtel and Graham (Bechtel, Abrahamsen, & Graham, 1998, 2001). Bechtel and Graham (op. cit.) rely on the authority of the cognitive psychologist George A. Miller so as to determine the birthdate of cognitive science as September 1956, at an MIT conference. Below, I will outline pieces of the history of cognitive science with a broad brush. This means that I will drop some names and allude to some ideas with an eye to their relevance to my own theory of the self in this book, but exclude some other approaches and ideas—for example, Turing computation, Gestalt psychology, and so on—which despite their historical importance do not bear on my own theory directly.

By the beginning of twentieth century, John B. Watson (1997), B. F. Skinner (1953), and a few others endeavoured to develop an experimental and anti-introspectionist approach to psychology—called behaviourism—which was allegedly in harmony with the objective methodology of natural sciences (Bechtel et al., 1998). There are behaviourist theories of various stripes, but we cannot here delve into a detailed review of the varieties of behaviourism (for a detailed review of the main tenets of various stripes of behaviourism, see Kantor 1968). Generally, behaviourists aimed to steer clear of hypothesising about the internal, mental mechanisms of the organism. Instead, they aimed to account for classes of correlations that specify the agent-environment relationship in terms of behavioural, non-mental, and physical factors. Because internal mechanisms of thinking are not publicly observable, behaviourists dismissed them and relied on behaviour—as the only kind of objective data—in hypothesising in psychology. Thus, behaviourists purported to reduce the internal mechanisms of thinking to classes of correlations between stimuli and responses.

From the perspective of the philosophy of science, behaviourism can be recognised as an instrumentalist view. It does not include any ontological commitments to mental entities or neuronal mechanisms that underpin them. In this book, I am developing a structural *realist* theory of the self. To the extent that the realist commitments are at issue, my project is the opposite of behaviourism and instrumentalism, because it embraces the kinds of commitments that are avoided by behaviourism (see my explanations about realism and structural realism in Chap. 2). The behaviourist approach to psychology has been subject to some well-known criticisms at any rate. One famous criticism holds that behaviourists' silence about the internal mechanisms of thinking borders on theoretical poverty. Later, during the 1940s, the development of information theory and cybernetics invoked psychologists to find viable theoretical tools for modelling internal mechanisms of thinking—in terms of information processing—and thereby overcome behaviourism's theoretical poverty (Bechtel et al., 1998, p. 6). It was within this context that the progress in cognitive psychology in works of Miller (1956) and Bruner (1956) broadened the scope of the science of cognition. Later, psychologists such as Miller lost interest in applying (Shannon's version of) information theory to their psychological enterprises and leaned towards Noam Chomsky's account of the grammatical structure of language—which was supposed to be an alternative to information theory according to Miller (2003, p. 141). Information, however, could be understood in a sense wider than what is insinuated in Shannon's mathematical theory of communication. Indeed, there are nice accounts of information processing in neural biological systems, which could be used to provide realistic models of internal mechanisms of cognition (Piccinini & Scarantino, 2011).

Be that as it may, the emphasis on the information theory and information processing is especially important from the perspective of our present enterprise because, in this book, I will specify the basic structure of the self in terms of information processing in the brain and nervous system. While there are different ways for characterising information processing, I am inclined to use Karl Friston's information-theoretic measure—characterised as the free energy principle (FEP)—to specify the basic informational structure of the self (see Chaps. 4 and 5 of this book). Thus, the bearing of information theory on the development of cognitive science is a significant historical point from the perspective of this enquiry. It is worth mentioning that FEP itself provides a unifying paradigm of cognitive neuroscience. FEP explains how it is that organisms that minimise the error in their cognitive models of the environment could maximise their survival. Part of my endeavour for fleshing out a structural realist account of the self relies on FEP. A comprehensive theory of the self needs to explain how it is that some organisms in such ecological circumstances could forge models that include self-conscious aspects, and FEP provides nice mathematical models of this model-making process. As I will explain later in this book, this unifying paradigm subsumes theoretical streams from AI and reinforcement learning, as well as information theory, and theoretical biology (Beni, 2018; Friston, 2010; Ramstead, Badcock, & Friston, 2017). Connectionism and reinforcement learning are among other theories that contributed to the development of contemporary cognitive science.

Connectionism was developed by McCulloch and Pitts (1943), Rumelhart and McClelland (1986), and Hebb (1949), among others. Psychological applications of connectionism presume that connectionist models could represent the activity of the brain's neurons and synapses via configuration of vector spaces in the connectionist networks. The approach presumes that the mode of information processing of the brain could be represented in terms of activation spaces embedded in the configurations of connectionist networks. The mode of the network's processing in connectionist networks—that is, parallel distributed processing (PDP)—is in contrast with classical computationalism according to which the central processing unit processes one discrete piece of the algorithm or program at a time.[7] It is believed that connectionist networks, their activation spaces, and their mode of processing (i.e., PDP) can provide a more realistic model of the brain's cognitive architecture, in comparison with what the classical version of computationalism—with a central processing unit—offers. The success of connectionism in providing viable models of internal mechanisms of thinking encouraged psychologists to go beyond the theoretical agnosticism of behaviourism about the internal mechanisms of thought, encouraging them to use connectionist models to represent the internal mechanism of thinking. Notice that even Alan Turing's classical computationalism (mentioned in the last footnote) assumes that it is possible to identify the brain's internal information processing in terms of digital computational mechanisms (Turing, 1950). However, even a mediocre knowledge of the brain's anatomy and internal functions reveals that classical computationalism does not provide a viable model of the biological brains' internal information processing mechanisms. This is because there are no experimental findings to suggest that the brain is a classical computer with a central processing unit and classical Turing

---

[7] By speaking of classical computationalism I am referring to digital information processing in a classical Turning machine. Turing's idea of computation is spelt out in terms of the processing in a universal Turing computer, which consists of a head that can read the symbols on an infinite type that is divided into discrete units (representing the steps of the algorithm). This abstract machine operates on the basis of a finite number of rules that indicate when the machine should write or erase symbols on the type, when it should move the type and proceed to the next instruction or halt. Turing conjectures that this abstract machine could process any computable algorithm (Turing, 1936).

computation mechanisms. Connectionism, on the other hand, provides artificial models that can indeed represent the internal structure of the information processing in the brain, with some amount of plausibility. This is because although neurology does not confirm the connectionist model completely, it can confirm that connectionism provides a viable model of distributed parallel processing in the brain. Below, I shall unpack this remark.

Since Ramon y Cajal's and Charles Sherrington's works on contiguous associations in the brain's structures (in the late nineteenth and early twentieth centuries), neurologists have had a good idea of the mechanisms of transmission of information which underpin the cognitive processes. Later, Hebb's (1949) theory of associative learning provided a more precise account of the brain's internal mechanisms of information processing. According to Hebb's theory, given the plasticity of the brain's neuronal population, the brain is capable of being trained by changing the configuration of the activation space of neural networks during the learning processes. The persistence or frequency of transmission of a message between two neurones (or neuronal populations) increases the synaptic efficiency between them. This is because neurons that are activated together frequently become inter-associated, in the sense that synaptic relations between them become strong. Mechanisms described by Hebb's theory of learning and its ingredients could be (and indeed have been) represented by connectionist networks. In this sense, connectionism provides new insights into the nature of information processing in the neuronal populations of actual brains. Connectionist models fill the blank that behaviourists left between stimuli and responses, explaining the brain's capacity for unsupervised learning, and explicating the nature of distributed information processing that underpins perception, memory, action, and other cognitive mechanisms.

In addition to the advancements in computer science and AI, the development of neuroimaging techniques enabled scientists to provide viable functional analyses of specific brain regions. The analysis of the brain functions on the basis of MRI, fMRI, PET, and similar techniques helps cognitive scientists to evaluate the veracity of their AI models of the brain's internal mechanisms; it allows them to provide more veridical models of the brain's information processing and internal mechanisms, without merely relying on introspection and verbal reports of the subjective modes. The internal mechanisms of cognition can be modelled by viable informational models, and the verisimilitude of such models can be ascertained on the basis of objective information produced by reliable neuroimaging techniques. Recent neurological developments, too, bear on our enterprise in this book. This is because my theory of the self in this book is based on neuroscientific insights into the function of specific parts of brain regions and mechanisms, for example, cortical midline structures (CMS), the Default Mode Network (DMN), the Mirror Neuron Systems (MNS), and their role in the processing of self-related information.

My philosophical account of personal and social aspects of the self draws on resources provided by recent advancements in theoretical and experimental neurology, the breakthroughs in computational models of the brain's information processing, and the experimental tools (e.g., neuroimaging techniques) that corroborate the viability of computational–informational models of thinking. The last remark about the computational–informational models needs some elaboration.

Although connectionism makes a great improvement on classical computationalism (in the sense of Turing computation), it does not provide a completely true model of the brain's information processing. Connectionist models are artificial (neural) networks, and although they specify the parallel and distributed nature of information processing in the brain correctly, there is no perfect match between the artificial network models and their less idealised and more complicated counterparts in the biological brains. In this book, I use "computation" and "information processing" in a wide sense that includes both classical computationalism and connectionism, as well as digital and analogue information processing, and is broader than all of these notions (for elaboration see Piccinini & Scarantino, 2011). One problem with this broad notion of computation is that it may be too permissive to provide a viable base for specifying the informational structure of the self. To narrow down the broad notion, I define the notion of computation as a sui generis form of information processing rendered by the cognitive mechanisms of an organism and through the organism's interaction with the environment. In submitting this definition, I follow Piccinini who defined computation in terms of the functional mechanisms of the physical system that performs concrete computation (Piccinini, 2007, 2015). I apply Piccinini's notion of physical computation to cognitive systems and set forth the notion of cognitive computation (in the sense defined a few lines ago). Please note that by making computation or information processing dependent on the cognitive mechanisms of the system, I somewhat consider the role of the embodied base of the information structure of the self. Instead of defining cognition in terms of abstract representations, embodied cognition theories emphasise the constative role of bodily mechanisms in cognition. In the same vein, enactivism emphasises the role of dynamical interaction or coupling between the organism and its environment in constitution of cognition (Di Paolo, 2014; Varela, Thompson, & Rosch, 1991). I will elaborate on embodied cognition and enactivism in the next chapters of this book. For the time being, it suffices to say that because I consider the embodied-ecological factors in their functional capacity (as mechanisms that perform the information processing) and because I do not eliminate computational models and representations completely, I am committed to a *moderate* form of embodiment and enactivism. My moderate approach is in line with previous research that cherishes the prospects of a meaningful dialogue between computationalism (or representationalism) and embodied theories (Pezzulo et al., 2011).

1.9   Concluding Remarks

The substantivalist theory of the self, as being advocated by venerable advocates such as Aristotle and Descartes among many others, possesses the status of orthodox theory. However, there are also philosophers who revolted against the orthodoxy; I mentioned the cases of Hume and Kant in this chapter and argued that it is possible to construe their views in terms of structuralist theory. The substantivalist view does not receive support from theories of contemporary cognitive sciences either. I have alluded to the origins of the emergence of cognitive sciences in the final sections of this chapter. Recent breakthroughs in empirical psychology can improve our philosophical understanding of the self, its nature, and its properties, without indicating that the self identifies with the orthodox substance.

The main issue that will be explored in the rest of the book is how to build upon recent scientific findings to provide a philosophically plausible account of the self without domesticating these findings to the orthodox substantivalist view.

My philosophical account of the self is centred on theories of experimental neuroscience, and more precisely, theories of information processing in CMS and MNS. Also, on account of being based on FEP, my theory of the self can receive support from evolutionary psychology and computational neurobiology. FEP draws connections between prediction error minimisation (PEM) mechanisms that the biological brain employs to model the causal structure of the world on the one hand and the evolutionary account of the organism's dynamical relationship with the environment and its homeostasis on the other. To remain in the state of equilibrium with the environment, the cognisant organism must be able to get a viable cognitive grip on the structure of the world. My theory of the self draws its biological plausibility from FEP. In all, I am developing a scientifically informed structuralist theory of the self which, although different from orthodox substantivalism, accomplishes a unifying account of different reflective, phenomenal, social, and ethical aspects of the self.

References

Aristotle. (1963). *Aristotle: Categories and De Interpretatione*. New York: Oxford University Press.

Aristotle. (2016). *De Anima* (C. J. Shields, Trans. & Ed.). Oxford: Oxford University Press.

Bechtel, W., Abrahamsen, A., & Graham, G. (1998). The Life of Cognitive Science. In W. Bechtel & G. Graham (Eds.), *A Companion to Cognitive Science* (pp. 1–104). Cornwall: John Wiley & Sons, Ltd. https://doi.org/10.1002/9781405164535.PART1

Bechtel, W., Abrahamsen, A., & Graham, G. (2001). Cognitive Science: History. *In International Encyclopedia of the Social and Behavioral Sciences*. https://doi.org/10.1016/B0-08-043076-7/01442-X

Beni, M. D. (2018). The Reward of Unification: A Realist Reading of the Predictive Processing Theory. *New Ideas in Psychology, 48*, 21–26. https://doi.org/10.1016/j.newideapsych.2017.10.001

Beni, M. D. (2019). *Cognitive Structural Realism: A Radical Solution to the Problem of Scientific Representation*. Cham: Springer Nature.

Brett, N. (1990). Hume's Causal Account of the Self. *Man and Nature, 9*, 23. https://doi.org/10.7202/1012607ar

Bruner, J. S. (1956). *A Study of Thinking*. New York: Wiley.

Descartes, R. (2008a). *A Discourse on the Method of Correctly Conducting One's Reason and Seeking Truth in the Sciences* (I. Maclean, Ed.). Oxford University Press. Retrieved from https://global.oup.com/academic/product/a-discourse-on-the-method-9780199540075?cc=ir&lang=en&

Descartes, R. (2008b). *Meditations on First Philosophy: With Selections from the Objections and Replies* (M. Moriarty, Ed.). Oxford University Press. Retrieved from https://global.oup.com/academic/product/meditations-on-first-philosophy-9780192806963?cc=ir&lang=en&

Di Paolo, E. A. (2014). The Worldly Constituents of Perceptual Presence. *Frontiers in Psychology, 5*(450). https://doi.org/10.3389/fpsyg.2014.00450

Fisher, S. (2005). *Pierre Gassendi's Philosophy and Science*. Brill. https://doi.org/10.1163/9789047416579

Friston, K. J. (2010). The Free-Energy Principle: A Unified Brain Theory? *Nature Reviews Neuroscience, 11*(2), 127–138. https://doi.org/10.1038/nrn2787

Graham, D. W. (2010). *The Texts of Early Greek Philosophy: The Complete Fragments and Selected Testimonies of the Major Presocratics*. Retrieved from https://www.cambridge.org/us/academic/subjects/classical-studies/ancient-philosophy/texts-early-greek-philosophy-complete-fragments-and-selected-testimonies-major-presocratics?format=WX&isbn=9780521845915

Hebb, D. O. (1949). *The Organization of Behavior: A Neuropsychological Theory*. New York: John Wiley and Sons. Retrieved from https://www.worldcat.org/title/organization-of-behavior-a-neuropsychological-theory-by-do-hebb-science-editions/oclc/3231198

Helmholtz, H. (1962). *Helmholtz's Treatise on Physiological Optics*. New York: Dover Publications. Retrieved from https://www.worldcat.org/title/helmholtzs-treatise-on-physiological-optics/oclc/523553

Hume, D. (2000). *A Treatise of Human Nature* (D. F. Norton & M. J. Norton, Eds.). Oxford University Press. Retrieved from https://global.oup.com/academic/product/a-treatise-of-human-nature-9780198751724?cc=ir&lang=en&

James, W. (1890). *The Principles of Psychology*. New York: Holt. Retrieved from https://archive.org/details/theprinciplesofp00jameuoft/page/n6

Kant, I. (1998). *Critique of Pure Reason* (P. Guyer & A. W. Wood, Eds.). Cambridge: Cambridge University Press. https://doi.org/10.1017/CBO9780511804649

Kantor, J. R. (1968). Behaviorism in the History of Psychology. *The Psychological Record, 18*(2), 151–166. https://doi.org/10.1007/BF03393755

McCulloch, W. S., & Pitts, W. (1943). A Logical Calculus of the Ideas Immanent in Nervous Activity. *The Bulletin of Mathematical Biophysics, 5*(4), 115–133. https://doi.org/10.1007/BF02478259

Melnick, A. (2010). *Kant's Theory of the Self.* New York: Routledge.

Miller, G. A. (1956). The Magical Number Seven Plus or Minus Two: Some Limits on Our Capacity for Processing Information. *Psychological Review, 63*(2), 81–97. Retrieved from http://www.ncbi.nlm.nih.gov/pubmed/13310704

Miller, G. A. (2003). The Cognitive Revolution: A Historical Perspective. *Trends in Cognitive Sciences, 7*(3), 141–144. Retrieved from http://www.ncbi.nlm.nih.gov/pubmed/12639696

Oppenheim, P., & Putnam, H. (1958). Unity of Science as a Working Hypothesis. In H. Feigl, M. Scriven, & G. Maxwell (Eds.), *Minnesota Studies in the Philosophy of Science* (Vol. II, pp. 3–36). Minneapolis, MN: University of Minnesota Press.

Pezzulo, G., Barsalou, L. W., Cangelosi, A., Fischer, M. H., McRae, K., & Spivey, M. J. (2011). The Mechanics of Embodiment: A Dialog on Embodiment and Computational Modeling. *Frontiers in Psychology, 2*(5). https://doi.org/10.3389/fpsyg.2011.00005

Piccinini, G. (2007). Computing Mechanisms*. *Philosophy of Science, 74*(4), 501–526. https://doi.org/10.1086/522851

Piccinini, G. (2015). *Physical Computation: A Mechanistic Account.* Oxford University Press. https://doi.org/10.1093/acprof:oso/9780199658855.001.0001

Piccinini, G., & Scarantino, A. (2011). Information Processing, Computation, and Cognition. *Journal of Biological Physics, 37*(1), 1–38. https://doi.org/10.1007/s10867-010-9195-3

Polansky, R. M. (2010). *Aristotle's De Anima: A Critical Commentary.* Cambridge University Press. Retrieved from https://www.cambridge.org/us/academic/subjects/classical-studies/ancient-philosophy/aristotles-de-anima-critical-commentary?format=PB

Ramstead, M. J. D., Badcock, P. B., & Friston, K. J. (2017). Answering Schrödinger's Question: A Free-Energy Formulation. *Physics of Life Reviews.* https://doi.org/10.1016/J.PLREV.2017.09.001

Rumelhart, D. E., & McClelland, J. L. (1986). *Parallel Distributed Processing: Explorations in the Microstructure of Cognition.* Cambridge, MA: MIT Press.

Shapiro, L. (2012). Cartesian Selves. In K. Detlefsen (Ed.), *Descartes' Meditations* (pp. 226–242). Cambridge: Cambridge University Press. https://doi.org/10.1017/CBO9781139030731.017

Skinner, B. F. (1953). *Science and Human Behavior.* New York: Macmillan. Retrieved from https://www.worldcat.org/title/science-and-human-behavior/oclc/191686

Strawson, P. F. (1959). *Individuals: An Essay in Descriptive Metaphysics.* London: Methuen.

Turing, A. (1936). On Computable Numbers, with an Application to the Entscheidungs problem. *Proceedings of the London Mathematical Society, 42*(3), 230–265.

Turing, A. (1950). Computing Machinery and Intelligence. *Mind, 59*(236), 433–460. https://doi.org/10.2307/2251299

Varela, F. J., Thompson, E., & Rosch, E. (1991). *The Embodied Mind: Cognitive Science and Human Experience.* Cambridge, MA: MIT Press.

Watson, J. B. (1997). *Behaviorism.* Routledge. Retrieved from https://www.routledge.com/Behaviorism-1st-Edition/Watson/p/book/9781351314329

Williams, B. (1978). *Descartes: The Project of Pure Enquiry.* Routledge. Retrieved from https://www.routledge.com/Descartes-The-Project-of-Pure-Enquiry/Williams/p/book/9781138019188

Wundt, W. (1980). *Outlines of Psychology* (pp. 179–195). Boston: Springer. https://doi.org/10.1007/978-1-4684-8340-6_7

Zoeller, G. (1993). Review Essay: Main Developments in Recent Scholarship on the Critique of Pure Reason. *Philosophy and Phenomenological Research, 53*(2), 445. https://doi.org/10.2307/2107782

2

Being Realist About Structures

The chapter outlines structural realism (SR), which is a flourishing theory of the philosophy of science. In this book, the structural realist theory of the self will be developed as a variety of SR. Given the importance of SR as a philosophical basis for this book's project, a rather detailed exposition of SR will be in order. In this chapter, I survey SR and explain that the ontic version of SR (OSR) comes with a revolutionary metaphysical component, which diverges from substantivalism and object-oriented metaphysics rather severely. According to substantivalism, what can be known (from the epistemological point of view) and what there is (from an ontological point of view) are to be specified as individual objects or substances. In contrast with substantivalism, SR advocates a form of scientifically informed philosophy according to which individual substances are not at the centre of philosophical attention. To be more precise, SR indicates that sciences provide viable structural descriptions of the world (according to epistemic SR), and the structure is all that there is (according to ontic SR). The chapter begins with a review of some persistent problems that motivated the emergence of SR in the field of the philosophy of physics, the notorious problems of pessimistic meta-induction (PMI) and metaphysical underdetermination.

© The Author(s) 2019
M. D. Beni, *Structuring the Self*, New Directions in Philosophy and Cognitive Science,
https://doi.org/10.1007/978-3-030-31102-5_2

As I argue, the object-oriented version of scientific realism cannot address these problems satisfactorily. SR, on the other hand, can provide viable solutions. I explain that SR has emerged with the promise of salvaging a modified version of scientific realism. It is the best of both worlds, in the sense that it can account for scientific progress in the face of PMI and the problem of metaphysical underdetermination, and it can do so without dismissing historical facts about theoretical changes (i.e., facts on which the antirealist draws her arguments). Neither does SR deny that there is a bewildering theoretical diversity in the foundations of modern physics. It acknowledges diversities but unifies them on the basis of commonalities and underpinning structures which lie at the foundations of SR semantics and metaphysics. After explicating the main tenets of SR, the chapter compares the situation in the fields of physics and psychology, with a focus on the problem of metaphysical underdetermination. Metaphysical underdetermination in the philosophy of physics was a result of the existence of diverse theoretical formulations of quantum statistics. I argue that there are similar cases in contemporary neuroscience that could cause a state of underdetermination of philosophical explanations by theoretical diversities in psychology. In the next chapter, I will build upon this claim to argue that under the circumstances, a structural realist theory of the self allows us to go beyond diversities and divergences and be structural realists about the underlying structure of the self.

This chapter also delves into some technical aspects of SR, arguing that SR overcomes pluralism by regimenting the representational commitments of the theories at the meta-level of abstract mathematical frameworks. In the same way, I argue that the meta-theoretical structure that regiments theories of selfhood can underlie the theoretic diversity of the field of cognition, and provide a guide for tailoring epistemological and ontological views to the scientific findings in the field of cognitive science and neuroscience.

2.1   Scientific Realism

As I explained in the last section of the previous chapter, behaviourists were professing agnosticism about cognitive mechanisms that relate stimuli to responses. However, progress in cognitive psychology not only launched a vehement quest for the discovery of such mechanisms, but it also motivated realism about scientific descriptions of cognitive mechanisms. Scientific progress provided a foothold for arguing that scientific theories, in the field of cognitive science and neuroscience, describe the internal cognitive mechanisms of thinking rather truthfully. Models of cognitive mechanisms described by scientific theories are at least approximately true and refer to real cognitive mechanisms in the brains and cognitive systems of actual human beings. We must face the issue of realism in cognitive science expansively in this book. But for now, I simply point out that scientific realism is already a well-established stance in the general philosophy of science and philosophy of physics. Below, I shall elaborate on the main themes and commitments of scientific realism.

Scientific realism makes commitments to a number of different philosophical theses across semantic, epistemic, and metaphysical or ontological domains. Let us begin with semantic commitments. A scientific realist accepts realist commitments to the scientific description of unobservable phenomena. In the philosophy of physics, a scientific realist assumes that theoretical descriptions of electrons, protons, or fields of energy are essentially veridical. In the philosophy of cognitive sciences, a realist presumes that theoretical accounts of the unobservable mechanisms of thinking are approximately true. The scientific realist not only presumes that the theoretical statements are approximately true, but she also asserts that knowledge conveyed by these true statements is truthful and reliable. Finally, the realist has to assert that the referents of the true scientific statements inhabit the real world and are factual states of affairs. A scientific realist cannot deny that theories are at least approximately true. Neither can she defend full-fledged realism without assuming that scientific knowledge is veridical and referents of theories exist in the real world. Stathis Psillos (2000b) has provided a viable formulation of three main theses of scientific realism; according to him, the renunciation of each one of these three theses leads to the endorsement of a version of antirealism.

Let us being with the semantic thesis. There is disagreement surrounding the referential credentials of observational terms. The real debate centres on the referential capacity of theoretical terms and their veracity. The semantic thesis of realism holds that scientific terms that refer to unobservable regions of the world have genuine truth-values. This means that theoretical terms of scientific theories (putatively) refer to factual unobservable entities that populate the world (Psillos, 2000b, p. 706). Versions of empiricism that deny that theoretical descriptions of the unobservable parts of the world are genuinely truth-conducive renounce the semantic thesis. An antirealist may even grant that we might assign truth-values to theoretical descriptions of unobservable regions of the world. However, she would argue that truth conditions of theoretical sentences should be reinterpreted in terms of (or be reduced to) the truth conditions of the observational sentences, that is, descriptions of observable phenomena. Such a reductionist approach has been attributed to various scientific empiricists such as Rudolf Carnap (Carnap, 1956; Psillos, 2000a). Allegedly, Carnap and his colleagues presume that theoretical statements have some "excess content", which cannot be directly explicated on the basis of the empirical content of scientific theories (Psillos, 2000b, p. 707). Carnap and cohort do not argue that we have to eliminate theoretical sentences from the scientific vocabulary. They grant that theoretical sentences contribute to systematising observations and organising them to result in new predictions. But because they deny that theoretical statements have genuine truth conditions independently of their role in systematising observations, they devised strategies—for example, such as Ramsey-sentence method—for accounting for truth-values of theoretical sentences by translating them into truth conditions of observational sentences (Beni, 2015). There may be a resemblance between the instrumentalist scepticism about the truth-values of theoretical terms and the behaviourist reluctance to engage in the theoretical discussion of unobservable mechanisms of thinking (for discussion, see the previous chapter, Sect. 1.8). Behaviourists, too, were averse to entertain a realist attitude towards theoretical models of unobservable internal mechanisms of thought. In the philosophy of cognitive sciences, a semantic realist will accept that the theoretical models of the unobservable mechanisms of thought have genuine truth-values. For example, she concedes that the statement "there is a causal relationship between depression and diachronic unity disorders" is genuinely truth-conditional. The realist does not assume that theoretical statements about the causal power of depression are only instrumental stipulations retained only on account of their usefulness for systematising what we know about the observable symptoms. According to the realist, the theoretical model may refer to real unobservable psychological facts, about real mechanisms of depression.

Be that as it may, a full-blooded version of scientific realism presumes that both theoretical and observational components of scientific discourse possess genuine truth-values. Note that scientific realists are not concerned with the strict truth or falseness of scientific theories but with their approximate truth. The hesitation in endorsing a strict notion of truth is due to the fact that scientific theories change and progress through time. Therefore, they seem to get closer to truth through changes and advancements. A strict conception of truth does not sit comfortably with a viable account of scientific progress, because a strictly true theory cannot change without becoming strictly false. Therefore, scientific realists usually speak of the approximate truth of mature scientific theories, which get closer to the truth as they evolve. In this vein, in scientific realist discourse, the notions of "truthlikeness" and "verisimilitude" take the place of strict truth or falsehood. And there are even probabilistic accounts of degrees of truth or verisimilitude of theories that support scientific realism (see Niiniluoto, 1987).

In addition to the semantic thesis, there is the epistemological thesis, which holds that approximately true theoretical descriptions of the world provide reliable knowledge of the nature and properties of entities that inhabit the world (Psillos, 2000b, p. 706). It is worth mentioning that there are some sophisticated versions of empiricism that do not reject the semantic thesis or metaphysical component—which will be explained shortly—but still reject the epistemic thesis of scientific realism. That is to say, some empiricists such as Bass van Fraassen (1980, 2002) grant that there might be a mind-independent world, and they assert that scientific statements are literally true, without also accepting that scientific theories can (or even aim to) provide knowledge about the unobservable regions of the world. A scientific realist, on the other hands, presumes that scientific descriptions provide genuine knowledge of the unobservable regions of the world, for example, knowledge of facts that could produce psychiatric disorders.

Finally, there is a metaphysical component to scientific realism. It holds that there is a mind-independent objective domain, that is, the world. The existence of the world, its structure, or the nature and properties of the kinds and entities that populate the world does not depend on our knowledge, recognition, or verification of what there is. The world's structure can be complicated, and parts of it are certainly unobservable. The unobservable regions, however, can be explored through experimentation, theorising, and scientific interventions. Thus, the scientific realist is committed to the metaphysical thesis which holds that the world (observable and unobservable regions included) is prior to and independent from our descriptions of the world. In contrast, a metaphysical anti-realist or agnostic may deny that we can make any commitments to the nature of mind-independent reality. The Kantian view which had been discussed in the previous chapter could motivate a modest form of scientific realism—for example, epistemic SR, which will be discussed later in this chapter. However, it compromises the metaphysical component of the full-blooded version of scientific realism. The Kantian view presumes that we cannot make any meaningful assertions about the structure or nature of entities that populate the world-in-itself. This means that we may remain agnostic about the objective domain in its full glory, which remains beyond the scope of our cognitive faculties and categories of understanding. Scientific realists, on the other hand, embrace the metaphysical thesis and commit themselves to the existence of the mind-independent objective domain. In the field of philosophy of cognitive science, a scientific realist may accept that there exist facts such as unobservable psychological entities, processes, and conceptual schemes that make mentalising and social cognition possible. She may accept that the self is a self-subsistent entity which exists prior to and independently from our theories and speculations.

I have to emphasise the point that a realist understanding of the notions of "truth" and "reference" brings three component theses of realism together. "Truth" and "reference" are primarily semantic notions. However, when a scientific realist asserts that descriptions of the unobservable regions of the worlds are essentially or approximately true, she means that theories refer to facts that inhabit a part of the world that exist independently from our recognition or attempts at determining truth-values. This means that the semantic component supports the metaphysical component (Boyd, 1983; Musgrave, 1992). Also, a scientific realist presumes that theories that are true and have veridical references to unobservable regions of the real world provide reliable knowledge of the unobservable parts of the world. Therefore, there is a close relationship between the realist understanding of notions of "truth" and "reference" on the one hand and the epistemic component of scientific realism on the other. Accordingly, the most famous argument for scientific realism aims to explain the empirical success of our best scientific theories on the basis of the verisimilitude of scientific theories. Because scientific theories are true and refer (or have veridical references) to unobservable parts of the world, scientific theories are instrumentally successful constructs. The connection between verisimilitude of theories and their empirical success is drawn by the famous "no miracle argument" (NMA), which, according to Hilary Putnam's formulation, holds that realism "is the only philosophy that doesn't make the success of science a miracle" (Putnam, 1975, p. 73). This is because, unless we assume that the empirical success of scientific theories is based on their truth and veridical references, we have to grant that the success of theories is due to a miracle, luck, or coincidence in cosmic scales (Psillos, 1999, p. 70). NMA can be understood as an explanationist defence of scientific realism, in the sense that it identifies the empirical success of scientific theories amounts with an explanandum, that is, a fact that demands explanation. In response to this call, the scientific realist submits that "The only reasonable explanation for the success of theories of which I am aware is that well-confirmed theories are conjunctions of well-confirmed, genuine statements and that the entities to which they refer in all probability exist" (Maxwell, 1962, p. 8; Psillos, 1999, p. 71). Maxwell supports his defence of realism by invoking probabilistic arguments and Bayesianism, to argue that the prior probability of the thesis of realism is higher than the prior probability of the thesis of instrumentalism[1] because the explanation that realism provides for the success of science is simpler and more comprehensive.

Regardless of the formulation of the argument (NMA, Bayesianism, etc.), the general insight behind the explanationist defences of realism is

---

[1] In the context of Maxwell's argument, the probability of the empirical success of science given the thesis of realism is equal to the probability of the empirical success of science given the thesis of instrumentalism. Therefore, the question boils down to the difference in the prior probability of realism and instrumentalism.

that the empirical success of scientific theories demands some kind of explanation, which cannot be offered by the antirealist. Presuming that empirical success of science is an explanandum, the scientific realist can provide a more plausible explanation for the success of scientific theories in comparison with her rival, that is, the antirealist, who ignores the demand for explanation or presumes that the success of theories is a result of coincidences in the cosmic scale.

It is also worth noting that the scientific realist could support the argument for the plausibility of realism on the basis of naturalistic views, by indicating that the explanationist defence of scientific realism is naturalistic because explanatory inferences are common enough in the scientific practice (Boyd, 1983; Putnam, 1975). Moreover, as Boyd argues the actual scientific practice and the application of the scientific methodology is heavily reliant on theoretical considerations (see Boyd, 1980, p. 618). Unless we can account for the reliability of the theoretical considerations that underpin scientific methodology and practice, even the empirical success of scientific theories cannot be justified. Thus, the explanatory defence of scientific realism is motivated by concerns about the rationality of scientific practice and methodologies, and it bears on scientific practice and the application of scientific methodology.

Let me recap. Scientific realism aims to consolidate the view that scientific theories are essentially true (semantic thesis), in the sense that the referents of scientific statements are the facts that populate the world (metaphysical thesis), and approximately true scientific theories provide reliable knowledge of the world (epistemic argument). As I have argued, notions of "truth" and "reference", when understood in the spirit of realism (e.g., correspondence to the mind-independent states of affairs), engraft metaphysical epistemic components on to the trunk of the full-blood version of scientific realism (which is committed to all three theses). To consolidate her views, a scientific realist may argue that the empirical success of scientific theories must be explained on the basis of their truth or veridical references. In reaction to this view, the antirealist conjures counterarguments to the effect that there is no meaningful relationship between the empirical success of scientific theories and their truth or verisimilitude. I shall explain this point in the next section.

2.2   Antirealism

2.2.1   Pessimistic Meta-Induction (PMI)

Pessimistic meta-induction (PMI), as being stated by Larry Laudan (1981), draws on consecutive facts about theoretical changes in the history of science to challenge the assumption of the existence of a meaningful relationship between the empirical success of scientific theories and the verisimilitude of their references. And there are various cases from the history of science—the phlogiston theory of combustion, the caloric theory of heat, the electromagnetic ether, among several others—which seem to provide a plea for the historical viability of PMI. As Psillos has epitomised the argument for PMI:

> The history of science is full of theories which at different times and for long periods had been empirically successful, and yet were shown to be false in the deep-structure claims they made about the world. It is similarly full of theoretical terms featuring in successful theories which do not refer. Therefore, by a simple (meta-)induction on scientific theories, our current successful theories are likely to be false (or, at any rate, are more likely to be false than true), and many or most of the theoretical terms featuring in them will turn out to be non-referential. Therefore, the empirical success of a theory provides no warrant for the claim that the theory is approximately true. There is no substantive retention at the theoretical, or deep-structural, level and no referential stability in theory-change. (Psillos, 1999, p. 96)

From surveying historical cases of false theories, which had been presumed to be true because they had been empirically adequate, PMI inductively infers that the present theories that are supposed to be true now (because they are empirical adequate) will turn out to be false later. The general conclusion of PMI is that the empirical success of scientific theories does not need to be based on their truth or verisimilitude, because empirically adequate theories usually turn out to be false.

As I remarked before, the semantic thesis of scientific realism is not based on the strict truth of theories so much as their approximate truth or verisimilitude. The scientific realist can ward off PMI's challenge by pointing out that although the history of science includes theories that had been strictly false, theories get closer to the truth as the scientific theories improve through history, and therefore, there is continuity underneath the theoretical changes that underlie PMI. Thus, PMI can be countered by a form of optimistic meta-induction which holds that although all expired theories are strictly speaking false, at least some central parts of mature theories approximate the truth. Through scientific progress, theories become less false. In the case of present theories, too, it can be claimed that the central theoretical parts of present theories are truer than previous theories. It follows that there is a connection between the empirical success of theories and their verisimilitude or approximate truth. Although this reply could account for theoretical continuity in the history of science and therefore support the realist account of the connection between the empirical success and veridical references of theories, it must provide a viable story of how to demarcate the central parts of the theories from their surplus or subsidiary parts (for discussion see Psillos, 1994). In my opinion, a more relevant question is how much of the precision of the notion of truth could be sacrificed without endangering the semantic and epistemological components of scientific realism. Let me elaborate.

There is a correspondence between the truth of theories and the resemblance that they bear to the states of affairs in the world. The realist may dispense with the strict truth of scientific theories, but then she should explain how it is that the resemblance of the theories to the world's states of affairs does not lose its significance when she replaces strict truth with approximate truth. Even scientific statements, say, about electromagnetic aether, could be approximately true. Even the theory of luminiferous aether can at least to some extent explain the wave-like propagation of light beams through space. This does not need to mean that basic concept of the theory, for example, "aether", refer to factual entities in the world, for example, aether. This means that the orthodox scientific realist's use of notions such as "approximate truth" does not help her to outmanoeuvre the antirealist's application of PMI. While scientific realists have suggested nice methods for measuring closeness to the truth (e.g., in Niiniluoto, 1987), the viability of the notion of approximate truth remains a moot point (see Worrall, 1989, p. 106).

2.2.2   The Issue of Underdetermination

The problem of underdetermination of theory by evidence poses a more serious threat (in comparison with PMI) to scientific realism. Versions of this problem have been stated by Pierre Duhem ([1906] 1954) and W. V. O. Quine (1953). I flesh out the problem immediately. The problem draws its force from confirmation holism, which holds that individual beliefs are not confirmed or disconfirmed by experience in isolation. Rather bulks of belief, for example, a scientific theory, or even a wide web of interrelated scientific theories and auxiliary hypotheses, are tested against evidence and receive confirmation or disconfirmation holistically. Given the soundness of antecedents of confirmation holism, the problem of underdetermination of theory by data holds that the same pieces of evidence might confirm diverse or even conflicting sets of theories. Thus, theories are underdetermined by data or evidence. Because it is not possible to confirm or disconfirm theoretical statements individually, the issue of underdetermination cannot be dissolved. As Bass van Fraassen has eloquently described the situation:

> The phenomena underdetermine the theory. There are in principle alternative developments of science, branching off from ours at every point in history with equal adequacy as models of the phenomena. Only angels could know these alternative sciences, though sometimes we dimly perceive their possibility. The theory in turn underdetermines the interpretation. Each scientific theory, caught in the amber at one definite historical stage of development and formalization, admits many different tenable interpretations. What is the world depicted by science? That is exactly the question we answer with an interpretation and the answer is not unique. Perhaps no interpretation ever finishes the task of answering all questions about the depicted world it displays as the theory's content. (Van Fraassen, 1991, pp. 480–481)

I shall unpack the bearing of the problem on the scientific realist stance. The scientific realist submits that scientific theories provide veridical representations of the world. The world, according to the realist thesis, is prior to our theoretical descriptions. However, our theories can describe the essential futures of reality and provide a guide to designing the ontology of science. Given the underdetermination of theories by evidence, the scientific realist cannot rationally choose the right set of theories (amongst empirically equivalent alternative sets) on which to rely for reaching a truthful scientific ontology. For each theoretical set that the realist chooses, there will be empirically equivalent rivals whose epistemic and ontological interpretations could be conceded with the same amount of rationality. This is because diverse sets of theories, with conflicting epistemological and ontological interpretations, can be recognized as true when put to the test against the same pieces of evidence.

Unlike PMI, the present form of the problem of underdetermination is not supported by concrete historical examples. Rather, it is based on general insights into the *conceivability* of empirically equivalent theories or interpretations. That is to say, the antirealist cannot substantiate the point that there are empirically equivalent rivals for a non-negligible number of scientific theories. Thus, in my opinion, the problem of underdetermination of theories by evidence does not pose an imminent danger to scientific realism. Moreover, as John Worrall has convincingly argued, two or several empirically equivalent rival theories must share at least the same observational components (because obviously, the problem indicates that theories could be equivalently verified by the same evidence!) (Worrall, 2011). However, the theoretical and observational components of scientific theories are hardly separable and theoretical considerations bear on observational implications of theories rather heavily. Theoretical and observational components of scientific theories are inseparable. Therefore, theoretical vocabularies of two or several empirically equivalent theories, which have the same observational components (because they are empirically equivalent!) would be at least partially overlapping too. This conclusion enables the realist to dissolve the problem. This is because the scientific realist can make ontological, semantic, and epistemic commitments on the basis of the overlapping, shared vocabularies of the presumably rival theories. Be that as it may, as I say, the problem does not pose a grave danger to scientific realism, I won't discuss it further in this book. However, there is another version of the underdetermination problem that deserves to be taken seriously. Below, I shall outline this alternative version of the problem.

The problem of metaphysical underdetermination is a variety of the underdetermination problem that poses a more serious threat to scientific realism. This is because the problem of metaphysical underdetermination draws on concrete cases from recent quantum statistics which indeed result in conflicting metaphysical consequences. As I will argue shortly, diverse theoretical discourses of quantum mechanics suggest completely different kinds of theoretical stipulations, which result in different kinds of ontological commitments. Let me elaborate.

The metaphysical underdetermination is not the same problem as the underdetermination of theories by the same evidence. The latter notion indicates that for any given theory there are empirically equivalent rival theories, whereas the former draws on specific instances of underdetermination of metaphysics by Quantum Mechanics (QM) (French, 2018). To be more precise, instead of relying on general insights into the possibility of empirically equivalent theories, the metaphysical underdetermination problem builds upon the implications of diverse formulations of non-relativistic many-particle quantum statistics to substantiate its claim. The claim is that because various formulations of Quantum Mechanics lead to various ontological packages which alternatively include and exclude individual objects, physics underdetermines metaphysics. This insight can be developed to indicate that scientific theories do not provide reliable knowledge of the real states of affairs in the world because scientific theories entail incompatible ontological consequences. I shall delve into details immediately.

The orthodox scientific realism is committed to a version of substantivalism or object-oriented ontology. For example, a realist in the field of physics may describe the "nature" of the entities that populate the world, for example, atoms, in terms of substances, individual objects, or entities with distinguishable properties, such as spatiotemporal locations (French & Ladyman, 2003, p. 35). However, as Steven French and Michael Redhead have argued science does not support this substantivalist ontological conceptions because theoretical terms of quantum physics, for example, "electrons", may refer to either individual or non-individual objects (French & Redhead, 1988). The situation leads to conflicting metaphysical packages with diverse commitments to the existence of individual objects (or lack thereof). As I will explain presently, French developed this theme on several later occasions.

In quantum statistics, there are two general ways of modelling the distribution of particles over energy states. Bose-Einstein statistics indicates that there are indistinguishable particles, such as bosons, over discrete energy states, whereas Fermi-Dirac statistics presumes that there are many identical particles, that is, fermions, with half-integer spins that are subject to the Pauli exclusion principle. Because the Pauli exclusion principle holds that particles with half-integer spins cannot occupy the same state within a system with thermodynamic equilibrium, it somehow supports the principle of identity of indiscernibles (PII, *aka* Leibniz' law).

PII is an ontological principle which holds that two (or more) entities that have completely identical properties and are completely indiscernible cannot be distinguished and therefore are identical. As French and Krause have argued, Fermi-Dirac statistics supports PII, because it holds that "the asymmetry of the wave function ensures that no two fermions can have the same intrinsic, or state-independent, properties, nor the same state-dependent properties expressed by expectation values of all quantum-mechanical physical magnitudes" (French & Krause, 2010, p. 150). On the other hand, there is the Bose-Einstein statistics, which indicate that bosons are indistinguishable and presume that they share all of their state-dependent properties. Consequently, Bose-Einsten statistics violates PII because bosons are not discernible (ibid.). Thus, the two models of the distribution of particles over states provide conflicting accounts of the existence of individual objects; Bose-Einstein statistics excludes the possibility of the existence of individual objects whereas Fermi-Dirac statistics recognises the existence of individual objects.

Notice that the same state of underdetermination would be observable if we focus on the distribution of particles over states, instead of the nature of particles. This is because classical statistics and quantum statistics endorse different accounts of the counting of the arrangement of particles over states. In a nutshell, classical statistics, that is, Maxwell-Boltzmann statistics, counts each permutation of the particles whereas, owing to its commitment to a form of symmetry (i.e., Permutation Invariance) quantum statistics, in the guise of both Bose-Einstein and Fermi-Dirac formulations, do not count permutation. I gloss over the scientific technicalities for the sake of simplicity (for technical elaboration see French, 2014, pp. 34–35; French & Krause, 2010). The take-home point is that, at least according to some interpretations, the fact that quantum statistics does not count permutation was regarded as an indicative of its renunciations of the notion of individuals. This is in contrast with the implications of classical Maxwell-Boltzmann statistics which is committed to the existence of atomic individual objects, and therefore counts a state and its permutation as two distinct states. Thus, the ontological implications of classical statistics and quantum statistics are incompatible.

It should be noted that even regardless of the ontological implications of classical statistics—which remains loyal to individual objects—diverse ontological implications of quantum statistics are not compatible with one another. For example, it is possible to assume that quantum statistics eliminates individuals from its ontology. Once this assumption is made, one can use quasi-set theoretical tools to describe the type of ontology that follows from the quantum statistical account of non-individual objects (French & Krause, 2010, chapters 7 and 8). On the other hand, it has been argued that quantum statistics retains the notion of individual objects. The argument for this claim is based on the action of Permutation Invariance which effectively divides the relevant Hilbert space into non-overlapping sub-spaces which represent permutation groups—including symmetric groups (bosons) and anti-symmetric groups (fermions). This alternative indicates that quantum entities can still be regarded as individuals. Their individuality, according to this reading, is a result of their being subject to certain constraints on their behaviour. To be more precise, the entities are constrained by certain sub-spaces of Hilbert space, given by the action of permutation invariance, and therefore they can be localized and individuated in terms of those sub-spaces (French, 2014, p. 47). Thus, the state of underdetermination between various conceptions of individual objects continues to hunt the ontological implications of quantum statistics.

Later, projecting this form of underdetermination to the scientifically informed conception of space-time points, Ladyman has argued that "[w]e need to recognise the failure of our best theories to determine even the most fundamental ontological characteristic of the purported entities they feature. It is an *ersatz* form of realism that recommends belief in the existence of entities that have such ambiguous metaphysical status"

(Ladyman, 1998, pp. 419–420). French and Ladyman's statement of the problem of metaphysical underdetermination is important from the perspective of our enterprise in this book because a similar situation in the field of cognitive science (and theories of selfhood) motivates my quest for constructing a structural realist theory of the self.

Let us recap. A scientific realist presumes that scientific theories provide truthful and veridical descriptions of non-observable individual objects that populate the real world. In this sense, the traditional scientific realism is object-oriented. Moreover, scientific realism is somewhat associated with naturalism, in the sense that it constructs its ontological thesis on the basis of what scientific theories reveal about the world. The metaphysical underdetermination problem draws on concrete examples from the recent quantum statistics to demonstrate that modern physics does not consistently ascribe the status of standard individual objects to particles. This would be a problem for the scientific realist because the ontological implications of modern physics are incompatible with the object-oriented conception of the nature of the entities that populate the world. As my brief review in this section indicates, the metaphysical underdetermination is more complicated than the underdetermination of theories by evidence. This is because it is reliant on concrete examples from theoretical resources of many particle physics—as being conceptualized alternatively in terms of quasi-set theory or by constraints in non-overlapping Hilbertian sub-spaces. Various formulations of quantum statistics entail inconsistent implications about the existential status of the individual objects. Thus, the problem of metaphysical underdetermination poses a serious challenge to orthodox scientific realism.

2.3   Structural Realism

Previously in this chapter (Sect. 2.2.1), I outlined PMI and explained how antirealists such as Laudan use its vicious force to deny that there is a meaningful relationship between the empirical success of theories and their truthfulness. A scientific realist cannot underestimate the force of PMI. As Psillos acknowledges, "[n]o realist can deny that Laudan's argument has *some* force. It shows that, on inductive grounds, the whole truth and nothing but the truth is unlikely to be had in science" (Psillos, 1999, p. 98). These words are chosen wisely by Psillos, because while scientific realists might concede that science cannot be "wholly" truthful, they proceed to argue that the approximate truth of mature scientific theories (which can be strictly false) is sufficient for establishing the realist claim about the verisimilitude of theoretical description of unobservable regions of the world. In this vein, the realist argues that an approximately true statement of the mature scientific theory represents the central features of the world faithfully enough (Psillos, 1999, p. 98). In the face of this optimism, there are various reasons to think that the notion of "approximate truth" is somewhat vague (for a formal assessment of the notion see Tichý 1974; for a philosophical evolution see Worrall 1989, pp. 104–106). And even if we could clarify the rather vague notion of "approximate truth" in terms of theories' capacity for representing the central features of the world, PMI can target the realist claim about the relation between central theoretical terms and central features of the world. That is to say, the scientific realist cannot demonstrate that central concepts of the present theories do represent central features of the world. Thus, it could be remarked that even in the case of the present theories, we cannot relate the empirical success of (our use of) some theoretical terms to their central status in the overall theoretical construct or to their veridical references to central features of the world. Structural realism has been originally developed to defend a modified version of scientific realism in the face of PMI, albeit without underestimating the historical facts about serious theoretical changes in the history of scientific progress.

Inspired by Henri Poincare's ideas about the importance of mathematical structures, John Worrall (1989) has endeavoured to present SR as a philosophical stance that recognises the historical evidence for theoretical changes. Despite this modest recognition of PMI's force, SR retains realist commitments. SR invokes a structuralist strategy to reconcile PMI's view on historical changes to realist commitments based on the verisimilitude of essential parts of shifting scientific theories in their specific domains (e.g., optics, electromagnetic, etc.). That is to say, according to SR, it is possible to specify the essential or central parts of theories structurally. It can be argued that there is structural continuity underpinning changes in theoretical content. To flesh out his claim, Worrall focuses on some controversial parts of the history of science, namely parts that could support PMI's account of theoretical changes, but at the same time shows that there is structural continuity underneath the changes. For example, he scrutinises the history of optics and the case of the theoretical shift from Fresnel's wave theory to Maxwell's theory of electromagnetic, and the shift from classical mechanics to relativity physics (Worrall, 1989, pp. 107–109). Worrall defers to the antirealist opinion that the observable theoretical shifts in the history of optics and physics are too radical to support any realist claim about the sameness of references of the central terms of the old and new theories. The historical changes indicate that theoretical concepts do not refer to the same entities, and the intrinsic nature of the theoretical entities cannot be specified satisfactorily in the face of theoretical modifications. However, by comparing the basic structure of the theories in the relevant fields, for example, electromagnetic, Worrall concludes that the continuity of theories is observable at the level of the form or structure of theories of a given field, not their content. Because there is theoretical continuity, there is cumulative development in various scientific domains. This indicates that science is in the business of consistently unveiling facts about various domains of the world, provided that we accept that the knowledge of the fact is structural, and science is not informative about the intrinsic nature of the entities that populate the world. Because Worrall's version of SR emphasises the structural nature of our *descriptions* of the world without committing itself ontologically to the (structural) nature of the world itself, it is usually named epistemic SR (ESR). There is also an ontic version of SR.

Ontic SR (OSR) is introduced by James Ladyman and developed by Steven French and others, and it comes with straightforward metaphysical commitments. It holds that structure is all that there is (Ladyman, 1998). While ESR emerges as a strategy that can address PMI, OSR aims to provide a solution to the problem of metaphysical underdetermination (mentioned in Sect. 2.2.2). As I remarked, orthodox versions of scientific realism cannot address the problem of metaphysical underdetermination, which is based on the concrete examples from the scientific practice, that is, models that violate the principle of identity of indiscernibles (PII).

Because the orthodox form of scientific realism is metaphysically committed to the existence of individual objects that populate the world, it cannot hold its ground in the field of fundamental physics after the violation of PII. Notice that, from the perspective of orthodox metaphysical theories, objects must be discernible to be individuated. This is in line with the insight behind the notorious Quinean aphorism which holds that there is no entity without identity (French, 2011a, p. 215). A substance must be discernible and distinguishable to be individuated. Without discernibility, there will be no substances, and without substances, orthodox metaphysics will lose its purchase. There might be forms of scientific realism that attempt to save the metaphysical commitments to individual objects in the field of many-particle physics by trying to domesticate scientific findings to the orthodox metaphysical views of substantivalist tendency. Such views might assume that, given our commonplace intuitions, individual objects really are constituting the world, but physics is not mature enough to be in harmony with the orthodox metaphysical insight.

Orthodox metaphysicians might argue that the fact that physics cannot find a handle on the intrinsic nature of those fundamental individual objects provides a reason for pessimism about the eligibility of physics for informing metaphysics. In this way, orthodox metaphysics might retract the naturalistic tendency of scientific realism and attempt to habituate scientific findings on the basis of stale metaphysical views. But the move would seem reactionary to the advocates of OSR (Ladyman & Ross, 2007). There might be forms of scientific realism that endorse agnosticism about the intrinsic nature of objects despite professing belief in the existence of individual objects in terms of substances, haecceities, or similar notions that are used for conceiving of fundamental units of reality. As Ladyman argues, "It is an *ersatz* form of realism that recommends belief in the existence of entities that have such ambiguous metaphysical status. What is required [in the face of the challenge of metaphysical underdetermination] is a shift to a different ontological basis altogether, one for which questions of individuality simply do not arise" (Ladyman, 1998, p. 420). OSR embarks on figuring out such alternative ontological basis. In order to face the challenge of metaphysical underdetermination, OSR dispenses with orthodox metaphysical prejudices and submits that it is possible to be realist not about the individual objects such as fermions or electrons, but about structural commonalities that are described by rival theoretical formulations that apply to the same domain. To be more precise, the solution that OSR offers consists in focusing on the group-theoretical structures that subsume diverse formulations of quantum statistics and using them to flesh out epistemological and metaphysical commitments, instead of trying to domesticate them to the discourse of the object-oriented, substantivalist ontology.

The structural realist solution to the problem of metaphysical underdetermination consists in dispensing with the substantivalist conception of entities as individual objects. Instead, OSR reconceptualises objects in structural terms (French, 2011a, p. 217). This means that instead of conceiving of objects as individual substances, SR conceptualises objects as structured entities that do not need to depend on an individual classical substance in order to exist. This proposal runs against the grain of the object-oriented scientific realism, according to which the unobservable individual objects such as electron are fundamental ontological units. Accordingly, the object-oriented realist presumes that relational properties—for example, charge, mass, and so on—that are transpired by the mathematical structure of the theory are worthwhile only because they serve to infer the hidden nature of unobservable individual objects—for example, electrons. The fact that physics does not support a substantivalist conception of particles is taken by the orthodox realist as a token of physics' unreliability as a guide to ontology. However, as the problem of metaphysical underdetermination indicates, there is no unique way of inferring the nature of the individual objects in many-particle physics. Under the circumstances, ontic structural realism (OSR) submits that the objects must be reconceptualised in structural terms, and structures themselves are fundamental (French, 2014, pp. 43–44). This brings us back to the central tenet of OSR (the eliminativist version of it), according to which the structure is all that there is.

There are diverse versions of ontic structural realism. There is an eliminativist version of OSR, according to which there are no such things as individual objects and there exists nothing but the structure. It holds that the structure is fundamental and the objects can be dispensed with altogether. French has supported this eliminativist version of OSR (French,

2014, 2018; French & Krause, 2010; French & Ladyman, 2003). The advocates of the eliminativist version of OSR have argued that there are scientific reasons—such as no-go theorems in Quantum Field Theory—which preclude the assumption of the existence of individual objects because the assumption harbours paradoxical conclusions (Muller, 2011). There is, however, also a non-eliminativist version of OSR which aims to retain an ontologically thin notion of individual objects. Non-eliminativist structuralists include a dwindled form of individual objects in their ontology, in addition to structures. Advocating non-eliminativist SR, Michael Esfeld argues that there are both individual objects and structures, but neither of them is ontologically primary, and the objects could be identified in virtue of their place in the structure (Esfeld & Lam, 2008). Similarly, in some elaborations, Ladyman argues that the assumption of the existence of (thin) individual objects is not *ipso facto* inconsistent with the negative tenet of OSR, according to which not all structural relations should be underpinned by self-subsistent individual objects with intrinsic distinguishable properties (Ladyman, 2007, 2018). Previously I have applied a version of non-eliminativist OSR to the field of psychology (Beni, 2018b) and developed a structural theory of selfhood (Beni, 2016b). I believe non-eliminativist OSR provides an adequate venue for fleshing out a structural realist theory of the self because, despite dispensing with the substantivalist conception of the selfhood, the non-eliminativist version of OSR allows for retaining the notion of weakly discernible individual selves. It may also allow for retaining weakly discernible non-structural aspects of the self, for example, the feeling of agency or ownership. This means that the non-eliminativist version of OSR possesses the requisite metaphysical resources to account for the self's identity over time (personal identity) and the phenomenal feeling of personhood. I will discuss this issue more expansively in the next chapters of this book.

2.4 The Significance of Informational Structures

There is also an informational version of SR (ISR). ISR has been developed in a number of different ways. ISR uses information-theoretic tools to flesh out its structural realist commitments across all three semantic, epistemic, and metaphysical aspects. In this respect, ISR is different from the Ladyman-French version of OSR, which regiments scientific representations in formal frameworks derived from set/model theory.

In the philosophy of science, there is a well-established tradition (dubbed the semantic view of theories, SVT) which holds that it is best to represent the structure of scientific theories in terms of set/model theory instead of, say, propositions or linguistic statements (Suppes, 1967; van Fraassen, 1969, 1980). Reasons for the dominance of the SVT in the philosophy of science of the late twentieth century are discussed extensively (Beni, 2019b, Sect. 2.3; French & Ladyman, 1999; Suppe, 1998). Here, I do not engage the discussion of the historical success of the SVT. Suffice to say that Ladyman and French have supplied their version of structural realism with the SVT's model-theoretic conception of theories because arguably the set/model-theoretic approach to theories wears its structural realist sympathies on its sleeves (Bueno & French, 2011; French & Ladyman, 1999; Ladyman, 1998). ISR is different to the set/model-theoretic version of OSR on account of the fact that firstly, ISR holds that structures are to be specified information-theoretically, and secondly, it makes epistemological and ontological commitments to informational structures. According to Floridi's statement of Informational Structural Realism:

> Explanatorily, instrumentally and predictively successful models (especially, but not only, those propounded by scientific theories) at a given LoA [i.e. Level of Abstraction] can be, in the best circumstances, increasingly informative about the relations that obtain between the (possibly subobservable) informational objects that constitute the system under investigation (through the observable phenomena). (Floridi, 2008, pp. 240–241)

While I can imagine that Floridi's version of ISR provides a suitable basis for fleshing out a structuralist theory of the self (Beni, 2016b), for reasons that are discussed elsewhere (Beni, 2016a, 2018d), I hold that Floridi's ISR leans towards epistemic structural realism. This means that I am sceptical of the metaphysical credentials of Floridi's version of ISR. The point about the epistemic leaning of Floridi's version is mainly for clarification and does not aim to indicate that ESR (or Floridi's version of ISR) is in any way inferior to OSR. Because the structural realist theory of the self should elaborate on the metaphysical aspects of the self, too, I do not think Floridi's version of ISR provides an adequate venue for developing it.

Ladyman and Ross (2007), too, developed a version of ISR, which is more straightforwardly committed to OSR, and presumes that structures can be characterised in terms of real informational patterns. Ladyman and Ross conjure the technical notions of projectibility and logical depth to specify the so-called real informational patterns. According to them, a pattern $x \to y$ is real if

1. It is projectible; and
2. It has a model that carries information about at least one pattern P in an encoding that has a logical depth less than the bitmap encoding of P, and where P is not projectible by a physically possible device computing information about another real pattern of lower logical depth than $x \to y$. (Ladyman & Ross, 2007, p. 233)

According to this view, the relation $x \to y$ is a projection from x to y, if it denotes any effected computation that yields y as output given x as input. More generally, $x \to y$ is projectible if, firstly, there is a physically possible computing machine that could perform the projection $x \to y$ given some scale of resolution, and secondly, if there is at least one other projection $x \to z$ that the computing machine can perform without changing its program (Beni, 2017b, p. 292; Ladyman & Ross, 2007, p. 224). Despite the reliance on notions of "projectibility" and "logical depth", the notion of information that is used by Ladyman and Ross' is not clearly defined. Nor do they argue why projectible patterns, as defined by them, should be the real patterns that are ontologically constitutive. This may raise doubts as regards the capacity of Ladyman and Ross' version of ISR for demarcating what they call real patterns from mere patterns or patterns simpliciter (see Beni, 2017b).

Be that as may, we need not delve into the technical details to reassess the merits or shortcomings of Floridi's or Ladyman and Ross' respective versions of Informational Structural Realism. Suffice it to say that, when it comes to presenting a structural realist theory of the self and its properties, it would seem more natural to try to specify the underlying structures in information-theoretic terms, rather than set/model-theoretic terms. This optimism about the efficiency of information-theoretic frameworks for fleshing out the structural realist theory of the selfhood is mainly due to the significant role of theories of information processing and computation in providing advanced models of the operation of the brain and cognitive system. The self and its properties are supposed to be mainly studied within the field of psychology and cognitive science. And as I explained in the final section of the previous chapter, cognitive science is a multidisciplinary venture with strong links to computer sciences and information theory, among other disciplines. As my selective references to the history of cognitive science in the previous chapter indicate, theories of computation and information processing bear strongly on our conception of how the brain and cognitive system work. This had been the case since Turing's epoch-making work on computation (Turing, 1950). The question of the relationship between the biological neural systems on the one hand and some specific forms of computation and information processing on the other still stirs a lot of scientific and philosophical curiosity. For example, Gualtiero Piccinini and colleagues have engaged in several interesting enterprises (Piccinini, 2015; Piccinini & Bahar, 2013; Piccinini & Scarantino, 2011) with the aim of specifying the types of information processing that are rendered by the physical and biological (also neural) information processing systems. The structural theory of the self which will be fleshed out in this book is mainly informed by scientific theories that identify the self in terms of information processing in specific regions of the brain, for example, cortical midline structures and Mirror Neuron Systems. In this light, it just seems natural to endeavour to flesh out the structural realist theory of the self along the lines of an information-theoretic version of SR.

On the same subject, it is important to notice that it is possible to conjure various senses of the notion of "information" and apply them to the construction of ISR. It seems that Floridi implicitly relies on a strongly semantic theory of information (Floridi, 2004) to supply his version of informational SR (for discussion see Beni, 2018c). Ladyman and Ross invoke the notions of projectibility and logical depth. (For technical elaboration see Ladyman and Ross 2007, pp. 222–237, and for discussion see Beni 2017b.) It is also worth noting that some of the ideas in Ladyman and Ross' informational OSR are inspired by John Collier's life-long engagement with the ideas of biological information and physics of information (Collier, 2003, 2011). But Collier was not an ontic realist, in the same way that Ladyman and Ross are. And while Collier himself has been one of the co-authors of the relevant chapter in Ladyman and Ross' book, the developed version of informational OSR is not totally loyal to Collier's views on the relationship between informational patterns on the one hand, and physical and biological structures in the real world on the other (as expressed in Collier, 2003, 2011).

More recently, I have constructed a version of SR which specifies informational structures in terms of embodied informational structures, that is, informational chains processed by and embedded in the neural informational processing mechanisms of the biological brains (Beni, 2018c, 2019b). In that enterprise, I have expansively substantiated the philosophical plausibility of a cognitive version of SR (CSR) on naturalistic and ecological grounds (Beni, 2019b, chapters 6 and 7). Despite praising the expressive power of set/model-theoretic frameworks or even abstract informational structures, I have argued that such formal tools are too abstract to substantiate the realist component of SR. As I have argued, the deep chasm that the orthodox versions of SR carve between the semantic component of OSR and its ontological component distract from the generation of a meaningful relationship between underpinning structures of scientific theories on the one hand and the causal structure of the world on the other. It is my view that if we make concession on regimenting scientific structures in terms of cognitive structures, that is, the informational structures implemented in the biological brains, we could address the question of scientific representation with remarkable naturalistic plausibility. This is because scientific theories are by-products of the activity of the biological brains of creatures such as ourselves, and there are indeed viable scientific accounts of how the brains of biological organisms could represent the causal structure of the world. The free energy principle and the Bayesian brain theory are among such neat scientific accounts (Friston, 2010; Friston & Stephan, 2007). CSR suggests that it is possible to build upon the scientific account of the brain-world representational relationship to account for the conundrum of scientific representation in the context of structural realist literature.

I do not think we need to spend more time reviewing details about various versions of SR and their merits and possible shortcomings. Such explications and assessments could be found in the expanding expository literature which fortunately includes some encyclopaedia entries authored by the founders of SR.[2] The point that must be taken into consideration with regard to the present enterprise is that a structural realist theory of the self will prosper in an information-theoretic venue. In my opinion, the structural realist theory of the self could flourish even more under a version of SR which specifies informational structures in terms of information processing in the biological brain and neural systems of organisms such as ourselves, or other groups of primates. I shall elaborate on this issue in later chapters of this book (especially Chaps. 4, 5, and 6). Thus, I end this section by asserting that cognitive SR seems to be specifically hospitable to a structural realist theory of the self, which specifies the basic structure of the self in terms of information processing in certain regions of the brain and nervous system of the organism.

2.5   The Level of Representation

There is another debate within the structural realist literature, whose discussion may shed some light on the philosophical characteristics of the structural realist theory of the self that I will flesh out in the next chapters of the book. The main point concerns the distinction that some structural realists make between the representational function of mathematical structures on the one hand and the ontological role of physical structures on the other. The issue becomes important in the context of the works of the advocates of set/model-theoretical approach to OSR. I will argue that it also bears on the present enterprise for developing a structural realist theory of self. Below, I shall unpack this point.

The point about the level of representation has emerged through a debate between Elain Landry (Brading & Landry, 2006; Landry, 2007) and Steven French (2011a, 2014). Defending a minimal version of SR,

---

[2] See entries authored by French and Ladyman in the *Stanford Encyclopaedia of Philosophy*: https://plato.stanford.edu/entries/qt-idind/; https://plato.stanford.edu/entries/structural-realism/

Landry argues that it is best to specify the "shared structure" of theoretical models of a theory at the level of scientific practice, and account for the commonality between structures in terms of the mentioned shared structures. According to this view, there is no need to characterise the shared structure set-theoretically, in order to show that two or several models have commonality or realise the same structure. It would be enough to show that "there is a morphism between the two systems, qua mathematical or physical models, that makes precise the claim that they share the appropriate kind of structure" (Brading & Landry, 2006, p. 2). According to this view, the relationship between scientific theories and the world has to be determined within various scientific contexts, and there is no telling that set theory is the universal formal framework that could be used to characterise the shared structure at the context of every scientific practice (Landry, 2007). This poses a serious challenge for the model-theoretical approach to ontic structural realism. In response to this challenge, French clarifies that he and other advocates of the model-theoretical approach to OSR do not presume that shared structures are characterised set-theoretically at the level of scientific practice (French, 2011a, p. 218). French's reply is based on the assumption of a distinction between structural presentations and representations. Structural representations of putative objects of theories at the level of scientific practice could be regimented group-theoretically, or on the basis of whatever mathematical language that is shared by the theories in question. On the other hand, there are representational structures which, as the advocates of the model-theoretic approach presume, can be formulated in terms of set/model theory. The point about the role of set/model theory can be confusing because it may indicate that the set-theoretical structures play an ontological role. However, as French clarifies, this construal would be wrong because ontic structural realists do not support a Pythagorean view—in the sense of making ontological commitments to set-theoretic entities (French, 2014, p. 10). Set-theoretical entities play a genuinely representational role. Physical structures, on the other hand, are ontologically constitutive.

Perhaps it is also worth mentioning that French and colleagues have expanded Tarski's model theory to provide a viable account of scientific representation through mathematical structures (Bueno, French, &

Ladyman, 2002; da Costa, Bueno, & French, 1998; da Costa & French, 2003). Here, the general idea is that model theory could be enriched—by elaborating on ideas of pragmatic truth and partial isomorphism or rather isomorphism between partial structures. The enrichment can take place with the explicit aim of enhancing the expressive power of the model-theoretic framework. After pragmatically enriching the quasi model-theoretic framework and enhancing its expressive power, the structural realists could use the formal framework to regiment inconsistencies that are included in scientific models in ways that remain beyond the scope of the orthodox model theory. I think the structural realists' engagement with the issue of the representational power of formal frameworks is both fruitful and liable to certain criticisms. Generally, I do not think that the set/model-theoretic framework—even the pragmatically enriched model-theoretic framework—provides the best possible venue of speaking of scientific representation (for discussion see Beni 2019b, Chap. 3, Sects. 3.5 and 3.6). More particularly, I do not think that set/model theory provides an adequate framework for regimenting the structuralist theory of the self. This is because, despite its great expressive power, model-theoretical formalism is not akin to the mathematical tools that are used in computational neuroscience.

For reasons mentioned in the previous section—for example, the reliance of this theory on scientific accounts of information processing in some specific brain regions—I assume that a structural realist theory of the self can prosper best within an information-theoretic framework. Thus, the information-theoretic framework can be preferred to the model-theoretic framework when it comes to developing a structural realist theory of the self. That being said, I have to add that despite being unsympathetic to set/model theoretical framework, I adhere to French's dichotomy between mathematical structures (which play a representational role) and physical structures which are ontologically constitutive. It is possible to adhere to this dichotomy without conceiving of scientific representations in terms of model-theoretic structures. This dichotomy finds its way into my informational-theoretic structural theory of the self. Let me elaborate.

The structural realist theory of the self that will be unfolded in this book identifies the underpinning structure of the selfhood in terms of patterns of information processing that are embodied primarily in certain areas of the brain, for example, cortical midline structures and Mirror Neuron Systems. It might be contended that understanding cognition or mindedness in terms of information processing implies reductionism about cognition. But please note that the non-eliminativist version of OSR is a modest form of realism that does not remove non-structural cognitive-mental aspects from ontology, given that this form of SR allows for retaining weakly discernible non-structural features of the self. Rather, the point is that scientific accounts of information processing in specific brain regions provide a good grip on the nature of self-structures that I use in constructing a scientifically informed metaphysical account of self-hood. Informational structures that specify the self are implemented in the brain regions, and by the same token, are associated with embodied aspects. On the same subject, please note that I am developing my structural realist account of the self on the basis of a specific form of informational SR, that is, cognitive SR (CSR), that has been presented on a previous occasion (i.e., Beni, 2019b). CSR specifies the underlying structure of theories in terms of embodied informational structures, that is, informational structures that are embodied in the information-processing mechanisms of the brain and cognitive system. Also, CSR presumes that to provide a reliable understanding of the environment (which is represented in scientific structures) the cognitive system couples with the causal structures in the world. I have previously delved into details to show how finding a cognitive grip in the world's windows of affordance (i.e., chances for action and cognition in the world) and coupling with it, maximises the organism's survival (Beni, 2019b, chapter 7). This means that CSR's account of underpinning cognitive-informational structures is associated with an embodied, ecological, and enactivist approach. I basically use the same notion, that is, "embodied informational structures", to develop my structural realist account of the self in this book. Thereby I submit that the informational patterns of the self could be extended to include the whole body, the environment, and the webs of social relations.

In the present enterprise, I embrace ontological commitments to the embodied informational structures that are primarily implemented in the brains and nervous systems but can also be extended to social structures and environment. These embodied structures can be represented mathematically sat in terms of (variational) Bayesian structures. Despite assuming that embodied informational structures are constituting the self (and thereby are ontologically constitutive), I concede that it would be useful to *represent* the structure of the self at a meta-level by invoking abstract informational structures. I need to speak more about predictive coding and the free energy principle before I can delineate the nature of abstract informational structures that represent the structure of the self at the meta-level of a Bayesian framework. I will do so in Chaps. 4 and 5 of this book, where I explain that the Bayesian framework could integrate various aspects of the self with outstanding unificatory power. However, for the time being, I assert that despite dispensing with set-model theory, my approach is loyal to French's dichotomy between representational and ontological structures specified as abstract informational structures versus embodied brain structures.

2.6   From Physics to Psychology and the Self-Concept

As my brief survey in Sect. 2.3 indicates, SR has its home in the philosophy of physics. Worrall's version of SR aims to address PMI by invoking examples from the history of optics, electromagnetics, and physics—that is, classical mechanics and relativity physics. French and Ladyman's version of SR (discussed in Sect. 2.3) aims to address the problem of underdetermination of metaphysics by quantum statistics. French's and Ladyman's works, too, are mainly concerned with philosophical problems raised by intriguing issues in the field of physics. Philosophical connotation of theories of physics, either in particle physics or relativity physics have motivated many other philosophical enterprises in SR literature (Lam & Esfeld, 2012; Muller, 2011; Pooley, 2006 among many others). Although SR has its home in the philosophy of physics, there have been interesting attempts at extending SR to the philosophy of special sciences. Among others, Elisabeth Lloyd (1994) and Steven French (2011b, 2013) endeavoured to provide structural realist theories of biology. There is no consensus about the right unit of natural selection in the philosophy of biology (Okasha, 2008), and the question of the real unit of natural selection in terms of, say, genes, individuals, groups, or species, could motivate a form of underdetermination. French argues that the structural realist strategy, that is, using set-theoretical structures to represent commonalities, could be re-enacted to shed some light on the debate about the right unit of selection, by saying that the underlying mathematical laws of evolution underpin the diversities and reconcile them. James Ladyman (2011) has argued for the merit of SR as a viable version of scientific realism that can account for scientific progress in the field of chemistry, by accounting for the structural continuity between theories of phlogiston and oxidation. Don Ross (2008) has defended an ontic structural realist theory of economics. And finally, I must remark that there have been attempts at extending SR into the field of cognitive sciences. For example, Hasselman, Seevinck, and Cox (2010) have endeavoured to show that a form of underdetermination in the field of cognitive sciences could be overcome by a structural realist ontology. Below, I paraphrase Hasselman et al.'s proposal.

There is a theoretical diversity in the field of cognitive sciences. An antirealist could build upon the observable theoretical diversity to argue that theories of cognitive science do not provide. Scientific practice in the field of cognitive science might support this antirealist claim because as I have argued in the previous chapter, cognitive science is a multidisciplinary enterprise. On such grounds, Dale and colleagues have argued that the plethora of coexisting (and sometimes rival) methods, hypotheses, and frameworks in cognitive science indicate that there is no unique way of explaining mental phenomena (Dale, 2008; Dale, Dietrich, & Chemero, 2009). According to this understanding of cognitive sciences, none of the rival theories, methods, and so on could be excluded.

The state of underdetermination of psychological explanations by diverse theories resembles the state of metaphysical underdetermination in the field of the philosophy of physics. In the field of philosophy of physics, various formulations of many particle physics endorsed conflicting ontological commitments. Similarly, it could be claimed that in the field of the philosophy of cognitive science, various methods, hypotheses, and theories invoke various explanatory frameworks. Both situations in physics and psychology include some underdetermination.

It is worth mentioning that Hasselman et al. do not see the situation in the cognitive sciences as an instance of underdetermination. They specify the problem of pluralism in terms of theory change and revisions in structure (Hasselman et al., 2010, p. 3). So, Hasselman et al. mainly speak about explanatory pluralism (rather than metaphysical underdetermination). However, OSR is primarily a theory of metaphysics. And I follow Steven French to consider underdetermination a metaphysical problem, rather than a problem of scientific explanations (although metaphysics and scientific explanations are somewhat relevant, see Psillos 2011). From the ontological point of view that I take in this book, the problem of underdetermination that wreaks havoc with the metaphysics of selfhood is ontological. Although I do not agree with their specification of the problem in terms of explanatory pluralism, I agree that as Hasselman et al. argue, it is possible to employ a structural realist strategy to address the issue of underdetermination (or theory change, according to their diagnosis) in cognitive science, and provide a measure for progress in this field. To do so, we have to presume that various research streams in cognitive sciences aim to discover the unifying structure beneath diversities and changes (Hasselman et al., 2010, p. 18). Aside from Hasselman et al., I have previously launched several enterprises for extending SR to the field of cognitive science. I have defended structural realist theories of the self, intentionality, and consciousness (Beni, 2016b, 2017a, 2018a, 2019a). In the same vein, Georg Northoff sought to apply SR to his theory of the brain's functions (Northoff, 2018). We do not need to delve into details to substantiate these proposals right now. Suffice it to say that there are cases of extension of SR to the field of cognitive science.

Let us recap. The take-home point is that the outbreak of various (at time conflicting) scientific theories of the self brings about a state of underdetermination of a metaphysical conception of the self by scientific theories of the self. In the next chapter, Chap. 3, I examine two scientifically informed theories of selfhood and show that the case of metaphysical underdetermination can indeed be reproduced in the field of the philosophy of self. After that, in Chap. 4, I show how a structural realist theory of the self enables us to overcome the state of metaphysical underdetermination in the field of cognitive science.

References

Beni, M. D. (2015). Structural Realism Without Metaphysics: Notes on Carnap's Measured Pragmatic Structural Realism. *Organon F, 22*(2015), 302–324.

Beni, M. D. (2016a). Epistemic Informational Structural Realism. *Minds and Machines, 26*(4), 323–339. https://doi.org/10.1007/s11023-016-9403-4

Beni, M. D. (2016b). Structural Realist Account of the Self. *Synthese, 193*(12), 3727–3740. https://doi.org/10.1007/s11229-016-1098-9

Beni, M. D. (2017a). On the Thinking Brains and Tinkering with the Scientific Models. *Axiomathes*, 1–15. https://doi.org/10.1007/s10516-017-9334-6

Beni, M. D. (2017b). Structural Realism, Metaphysical Unification, and the Ontology and Epistemology of Patterns. *International Studies in the Philosophy of Science, 31*(3), 285–300. https://doi.org/10.1080/02698595.2018.1463691

Beni, M. D. (2018a). A Structuralist Defence of the Integrated Information Theory of Consciousness. *Journal of Consciousness Studies, 25*(9–10), 75–98. Retrieved from http://www.ingentaconnect.com/contentone/imp/jcs/2018/00000025/f0020009/art00003

Beni, M. D. (2018b). Much Ado About Nothing: Toward a Structural Realist Theory of Intentionality. *Axiomathes.* https://doi.org/10.1007/s10516-018-9372-8

Beni, M. D. (2018c). Syntactical Informational Structural Realism. *Minds and Machines, 28*(4), 623–643. https://doi.org/10.1007/s11023-018-9463-8

Beni, M. D. (2018d). The Downward Path to Epistemic Informational Structural Realism. *Acta Analytica, 33*(2), 181–197. https://doi.org/10.1007/s12136-017-0333-4

Beni, M. D. (2019a). An Outline of a Unified Theory of the Relational Self: Grounding the Self in the Manifold of Interpersonal Relations. *Phenomenology and the Cognitive Sciences, 18*(3), 473–491. https://doi.org/10.1007/s11097-018-9587-6

Beni, M. D. (2019b). *Cognitive Structural Realism: A Radical Solution to the Problem of Scientific Representation.* Cham: Springer Nature.

Boyd, R. N. (1980). Scientific Realism and Naturalistic Epistemology. *PSA: Proceedings of the Biennial Meeting of the Philosophy of Science Association.* The University of Chicago Press Philosophy of Science Association. https://doi.org/10.2307/192615

Boyd, R. N. (1983). On the Current Status of the Issue of Scientific Realism. In *Methodology, Epistemology, and Philosophy of Science* (pp. 45–90). Dordrecht: Springer. https://doi.org/10.1007/978-94-015-7676-5_3

Brading, K., & Landry, E. M. (2006). Scientific Structuralism: Presentation and Representation. *Philosophy of Science, 73*(5), 571–581. https://doi.org/10.1086/518327

Bueno, O., & French, S. (2011). How Theories Represent. *The British Journal for the Philosophy of Science, 62*(4), 857–894. https://doi.org/10.1093/bjps/axr010

Bueno, O., French, S., & Ladyman, J. (2002). On Representing the Relationship Between the Mathematical and the Empirical. *Philosophy of Science, 69*, 497–518.

Carnap, R. (1956). The Methodological Character of Theoretical Concepts. *Minnesota Studies in the Philosophy of Science, 1*, 38–76. https://doi.org/10.2307/2964350

Collier, J. (2003). Entropy Hierarchical Dynamical Information Systems with a Focus on Biology. *Entropy, 5*, 100–124. Retrieved from www.mdpi.org/entropy

Collier, J. (2011). Information, Causation and Computation. *Of Foundations of Information and Computation, 2*, 89.

da Costa, N. C. A., Bueno, O., & French, S. (1998). The Logic of Pragmatic Truth. *Journal of Philosophical Logic, 27*(6), 603–620. https://doi.org/10.1023/A:1004304228785

da Costa, N. C. A., & French, S. (2003). *Science and Partial Truth*. New York: Oxford University Press. https://doi.org/10.1093/019515651X.001.0001

Dale, R. (2008). The Possibility of a Pluralist Cognitive Science. *Journal of Experimental & Theoretical Artificial Intelligence, 20*(3), 155–179. https://doi.org/10.1080/09528130802319078

Dale, R., Dietrich, E., & Chemero, A. (2009). Explanatory Pluralism in Cognitive Science. *Cognitive Science, 33*(5), 739–742. https://doi.org/10.1111/j.1551-6709.2009.01042.x

Duhem, P. M. M. (1954). *The Aim and Structure of Physical Theory*. Princeton University Press. Retrieved from https://press.princeton.edu/titles/2667.html

Esfeld, M., & Lam, V. (2008). Moderate Structural Realism About Space-Time. *Synthese, 160*(1), 27–46. https://doi.org/10.1007/s11229-006-9076-2

Floridi, L. (2004). Outline of a Theory of Strongly Semantic Information. *Minds and Machines, 14*(2), 197–221. https://doi.org/10.1023/B:MIND.0000021684.50925.c9

Floridi, L. (2008). A Defence of Informational Structural Realism. *Synthese, 161*(2), 219–253. https://doi.org/10.1007/s11229-007-9163-z

French, S. (2011a). Metaphysical Underdetermination: Why Worry? *Synthese, 180*(2), 205–221. https://doi.org/10.1007/s11229-009-9598-5

French, S. (2011b). Shifting to Structures in Physics and Biology: A Prophylactic for Promiscuous Realism. *Studies in History and Philosophy of Science Part C: Studies in History and Philosophy of Biological and Biomedical Sciences, 42*(2), 164–173. https://doi.org/10.1016/J.SHPSC.2010.11.023

French, S. (2013). Eschewing Entities: Outlining a Biology Based Form of Structural Realism. In *EPSA11 Perspectives and Foundational Problems in Philosophy of Science* (pp. 371–381). Cham: Springer International Publishing. https://doi.org/10.1007/978-3-319-01306-0_30

French, S. (2014). *The Structure of the World: Metaphysics and Representation.* Oxford: Oxford University Press. https://doi.org/10.1093/acprof:oso/9780199684847.001.0001

French, S. (2018). Defending Eliminative Structuralism and a Whole Lot More (or Less). *Studies in History and Philosophy of Science Part A.* https://doi.org/10.1016/J.SHPSA.2018.12.007

French, S., & Krause, D. (2010). *Identity in Physics: A Historical, Philosophical, and Formal Analysis.* Oxford University Press. Retrieved from https://global.oup.com/academic/product/identity-in-physics-9780199575633?cc=us&lang=en&

French, S., & Ladyman, J. (1999). Reinflating the Semantic Approach. *International Studies in the Philosophy of Science, 13*(2), 103–121. https://doi.org/10.1080/02698599908573612

French, S., & Ladyman, J. (2003). Remodelling Structural Realism: Quantum Physics and the Metaphysics of Structure. *Synthese, 136*(1), 31–56. https://doi.org/10.1023/A:1024156116636

French, S., & Redhead, M. (1988). Quantum Physics and the Identity of Indiscernibles. *The British Journal for the Philosophy of Science, 39*(2), 233–246. https://doi.org/10.1093/bjps/39.2.233

Friston, K. J. (2010). The Free-Energy Principle: A Unified Brain Theory? *Nature Reviews Neuroscience, 11*(2), 127–138. https://doi.org/10.1038/nrn2787

Friston, K. J., & Stephan, K. E. (2007). Free-Energy and the Brain. *Synthese, 159*(3), 417–458. https://doi.org/10.1007/s11229-007-9237-y

Hasselman, F., Seevinck, M. P., & Cox, R. F. A. (2010). Caught in the Undertow: There Is Structure Beneath the Ontic Stream. *SSRN Electronic Journal.* https://doi.org/10.2139/ssrn.2553223

Ladyman, J. (1998). What Is Structural Realism? *Studies in History and Philosophy of Science Part A, 29*(3), 409–424. https://doi.org/10.1016/S0039-3681(98)80129-5

Ladyman, J. (2007). On the Identity and Diversity of Objects in a Structure. *Proceedings of the Aristotelian Society, Supplementary Volumes.* Oxford University Press The Aristotelian Society. https://doi.org/10.2307/20619100

Ladyman, J. (2011). Structural Realism Versus Standard Scientific Realism: The Case of Phlogiston and Dephlogisticated Air. *Synthese, 180*(2), 87–101. https://doi.org/10.1007/s11229-009-9607-8

Ladyman, J. (2018). Introduction: Structuralists of the World Unite. *Studies in History and Philosophy of Science Part A.* https://doi.org/10.1016/J.SHPSA.2018.12.004

Ladyman, J., & Ross, D. (2007). *Every Thing Must Go.* Oxford: Oxford University Press. https://doi.org/10.1093/acprof:oso/9780199276196.001.0001

Lam, V., & Esfeld, M. (2012). The Structural Metaphysics of Quantum Theory and General Relativity. *Journal for General Philosophy of Science, 43*(2), 243–258. https://doi.org/10.1007/s10838-012-9197-x

Landry, E. M. (2007). Shared Structure Need Not Be Shared Set-Structure. *Synthese, 158*(1), 1–17. https://doi.org/10.1007/s11229-006-9047-7

Laudan, L. (1981). A Confutation of Convergent Realism. *Philosophy of Science, 48*(1), 19–49. https://doi.org/10.1086/288975

Lloyd, E. A. (1994). *The Structure and Confirmation of Evolutionary theory.* Princeton, NJ: Princeton University Press.

Maxwell, G. (1962). The Ontological Status of Theoretical Entities. In H. Feigl & G. Maxwell (Eds.), *Scientific Explanation, Space, and Time: Minnesota Studies in the Philosophy of Science* (pp. 181–192). Minneapolis, MN: University of Minnesota Press. Retrieved from https://philpapers.org/rec/MAXTOS

Muller, F. A. (2011). Withering Away, Weakly. *Synthese, 180*(2), 223–233. https://doi.org/10.1007/s11229-009-9609-6

Musgrave, A. (1992). Realism About What? *Philosophy of Science, 59*(4), 691–697. https://doi.org/10.2307/188137

Niiniluoto, I. (1987). *Truthlikeness.* Dordrecht: Springer. https://doi.org/10.1007/978-94-009-3739-0

Northoff, G. (2018). *The Spontaneous Brain: From the Mind-Body to the World-Brain Problem.* Cambridge, MA: MIT Press. Retrieved from https://mitpress.mit.edu/books/spontaneous-brain

Okasha, S. (2008). *Evolution and the Levels of Selection.* Clarendon Press. Retrieved from https://global.oup.com/academic/product/evolution-and-the-levels-of-selection-9780199556717?cc=us&lang=en&

Piccinini, G. (2015). *Physical Computation: A Mechanistic Account.* Oxford University Press. https://doi.org/10.1093/acprof:oso/9780199658855.001.0001

Piccinini, G., & Bahar, S. (2013). Neural Computation and the Computational Theory of Cognition. *Cognitive Science, 37*(3), 453–488. https://doi.org/10.1111/cogs.12012

Piccinini, G., & Scarantino, A. (2011). Information Processing, Computation, and Cognition. *Journal of Biological Physics, 37*(1), 1–38. https://doi.org/10.1007/s10867-010-9195-3

Pooley, O. (2006). Points, Particles and Structural Realism. In D. Rickles, S. French, & J. Saatsi (Eds.), *The Structural Foundations of Quantum Gravity* (pp. 83–120). Oxford: Oxford University Press.

Psillos, S. (1994). A Philosophical Study of the Transition from the Caloric Theory of Heat to Thermodynamics: Resisting the Pessimistic Meta-Induction. *Studies in History and Philosophy of Science Part A, 25*(2), 159–190. https://doi.org/10.1016/0039-3681(94)90026-4

Psillos, S. (1999). *Scientific Realism: How Science Tracks Truth*. New York: Routledge.

Psillos, S. (2000a). Carnap, the Ramsey-Sentence and Realistic Empiricism. *Erkenntnis, 52*(2), 253–279. https://doi.org/10.1023/A:1005589117656

Psillos, S. (2000b). The Present State of the Scientific Realism Debate. *The British Journal for the Philosophy of Science, 51*(4), 705–728. https://doi.org/10.1093/bjps/51.4.705

Psillos, S. (2011). Choosing the Realist Framework. *Synthese*. Springer. https://doi.org/10.2307/41477558

Putnam, H. (1975). *Mathematics, Matter, and Method*. Cambridge: Cambridge University Press.

Quine, W. V. O. (1953). Two Dogmas of Empiricism. In *From a Logical Point of View* (pp. 20–46). Cambridge, MA: Harvard University Press.

Ross, D. (2008). Ontic Structural Realism and Economics. *Philosophy of Science, 75*(5), 732–743. https://doi.org/10.1086/594518

Suppe, F. (1998). Understanding Scientific Theories: An Assessment of Developments, 1969–1998. *Philosophy of Science Biennial Meetings of the Philosophy of Science Association. Part II: Symposia Papers, 67*, 102–115. Retrieved from http://www.jstor.org/stable/188661

Suppes, P. (1967). What Is a Scientific Theory? In S. Morgenbesser (Ed.), *Philosophy of Science Today* (pp. 55–67). New York: Basic Books. Retrieved from https://www.google.com/_/chrome/newtab?espv=2&ie=UTF-8

Tichý, P. (1974). On Popper's Definitions of Verisimilitude. *The British Journal for the Philosophy of Science, 25*(2), 155–160. https://doi.org/10.1093/bjps/25.2.155

Turing, A. (1950). Computing Machinery and Intelligence. *Mind, 59*(236), 433–460. https://doi.org/10.2307/2251299

van Fraassen, B. C. (1969). Meaning Relations and Modalities. *Noûs, 3*(2), 155. https://doi.org/10.2307/2216262

van Fraassen, B. C. (1980). *The Scientific Image.* Oxford University Press. https://doi.org/10.1093/0198244274.001.0001

Van Fraassen, B. C. (1991). *Quantum Mechanics: An Empiricist View.* Oxford: Clarendon Press.

Van Fraassen, B. C. (2002). *The Empirical Stance.* Yale University Press. Retrieved from https://yalebooks.yale.edu/book/9780300103069/empirical-stance

Worrall, J. (1989). Structural Realism: The Best of Both Worlds? *Dialectica, 43*(1–2), 99–124. https://doi.org/10.1111/j.1746-8361.1989.tb00933.x

Worrall, J. (2011). Underdetermination, Realism and Empirical Equivalence. *Synthese, 180*(2), 157–172. https://doi.org/10.1007/s11229-009-9599-4

3

To Be Many or Not to Be, Grounds for a Structural Realist Account of the Self

In the previous chapter, I surveyed SR in the fields of the philosophy of physics and general philosophy of science. I also explained how various formulations of many particle physics lead to conflicting ontological consequences with regards to the existence of individual objects at the sub-particle level. This is bad news for the scientific realist because it indicates that scientific theories do not provide a viable description of the principal features of reality. The scientific realist cannot construct a scientifically informed ontology that reconciles the apparently conflicting assertions about the existence/non-existence of individual objects. In this context, ontic structural realists such as French and Ladyman offer to defend a modified version of scientific realism. To do so, they suggest that ontological commitments are to be made not with regard to individual objects, whose ontological status is marred, but to underpinning common structures that lie beneath various formulations of apparently conflicting

Parts of this chapter are reprinted with the kind permission from *Springer Nature*. The extracts are taken from two following papers of mine: 2016. "Structural Realist Account of the Self." Synthese 193 (12). Springer Netherlands: 3727–3740; 2018. "An Outline of a Unified Theory of the Relational Self: Grounding the Self in the Manifold of Interpersonal Relations." *Phenomenology and the Cognitive Sciences* 1–19.

© The Author(s) 2019
M. D. Beni, *Structuring the Self*, New Directions in Philosophy and Cognitive Science,
https://doi.org/10.1007/978-3-030-31102-5_3

theories. I also briefly explained that theoretical diversity in the field of cognitive science could lead to a situation similar to what is called metaphysical underdetermination in the field of physics. This chapter continues the course that was set in the previous one, focusing on a specific case of theoretical diversity in the field of cognitive science. I will argue that, in the field of neuroscience, various theoretical and experimental engagements with the self result in various neurophilosophical accounts of the self. Thus, the chapter argues that the same situation (i.e., metaphysical underdetermination) that motivated SR in the field of the philosophy of physics raises its head in the field of the philosophy of cognitive science and more specifically in the ontological account of the self.

The chapter canvasses the pluralist and the eliminativist theories of the self. It also outlines Gallese's and Northoff's respective theories. Sometimes these theories come with conflicting ontological implications. For example, the pluralist theory, as being represented by Shaun Gallagher's pattern theory in this chapter, indicates that the self consists of various patterns—including the narrative self, minimal self, semiotic self, social self, autobiographical self, an do on, whereas Thomas Metzinger's eliminativism indicates that there are no such things as selves at all. In all, I think the kinds of theories of the self that I have chosen in this chapter provide a viable idea of the confusing diversity of the neuroscientific and neurophilosophical accounts of selfhood. It is important to notice that none of these theories are in line with the substantivalist conception of the self (the substantivalist view explained in the first chapter of this book). Moreover, despite their relative empirical adequacy and theoretical strength, none of these theories provide an exclusively complete account of the self. Each one of these theories has some plausible features that make it recommendable, but it also has shortcomings that make it liable to objections. In this vein, I show that the diversity of theoretical descriptions of the self gives rise to a case of metaphysical underdetermination. After substantiating this point, the chapter suggests that we can use the same structural realist strategy that has been used in the philosophy of physics (as explained in the previous chapter) to face the challenge of underdetermination of ontological conception of the self with the diversity of the relevant theories in the field of cognitive science. According to this suggestion, ontological commitments have to be made on the basis of the commonality that could be discerned between diverse theories (or the common underpinning structure) that aim to describe the main features of the self. This paves the way for developing a structural realist theory of self in the next chapter.

3.1 Substantivalism and Its Demise

In the first chapter of this book, I sketched substantivalism as an intuitively appealing theory of the self. I suspect that most people think of their selves as enduring and somewhat independent entities that bear various properties. This common sense insight is in line with (and perhaps has been motivated by) some philosophical accounts of physical objects and selves as substances. As I explained in Chap. 1, some notable philosophers have advocated substantivalist accounts of the self, such as Aristotle, who introduced a substantivalist account of the self within the context of his hylomorphism, and Descartes, who endeavoured to argue for the plausibility of a different version of the substantivalist account of the self. When criticising the Cartesian argument for the existence of the self as an object-in-itself that endures through changes and preserves its identity over time, Immanuel Kant specified the self in terms of the transcendental condition of having experiences. In my view, it is best to understand Kant's theory of the self along the lines of structuralism (see Sect. 1.6 in the first chapter), although it is also possible to assume that Kant's theory is still committed to (some modified form of) substantivalism. Some contemporary philosophers have advocated this substantivalist construal of Kant's theory of the self. P. F. Strawson (1959), for example, revisited the substantivalist view in his interesting work on individuals, where he suggests that persons are some form of (Kantian) substances. Below, I briefly review the substantivalist view once more, albeit without delving into the fine historical details offered in the first chapter. Here, my aim is to set substantivalism as a foil, to show how the scientific conceptions of the self will diverge from substantivalism.

For the sake of argument, let us take a realist view of the world and the things that populate it. A number of classical philosophers, for example, Aristotle, Descartes, as well as Kant and Strawson, have adopted such a view. There can be (and indeed is) a division of opinion between philosophers who assume the things that populate the world are objects-in-themselves, and those who presume that we can know things only as they appear to us. However, both camps could agree that individual objects inhabit the world regardless of their view on the nature of the world. This means that, despite some subtle disagreements on how to develop realism, the substantivalist approaches agree that substances are the basic ontological units of reality. The world is the totality of trees, chairs, stones, and horses, and so on, which are basic ontological units. Some metaphysical views may also presume that the world includes treehood, chairhood, and so on as secondary substances. These units are somewhat independent and indivisible. Of course, it is possible to chop a tree into pieces. But the pieces are not individual objects. They are parts of the tree, and they still could be identified in virtue of their connection with the tree. A substance is also rather independent of other things in the world—assuming, for the sake of the argument that the sense of relative independence is clear in this context—and it retains some form of unity or identity over time. The substance of a tree, for example, cannot be reconstructed in terms of its parts, but the parts are identifiable in terms of their relationship with the unified substance which endures over time. The substance is the bearer of the properties (the tree is green, dry, tangled, etc.,) and it endures through the changes (grows, sprouts, gets dry, etc.). As I have argued in the first chapter, in addition to all kinds of physical things, selves can be recognised as primary substances too. It could be assumed that the selves are ontological units that constitute some regions of the world. A version of this perennial view—also manifested in works of Aristotle and Descartes—is defended by P. F. Strawson. Strawson (1959) argues that physical bodies and persons (or selves) are two fundamental types of particulars (or primary substances) that constitute the world. The self is an indivisible entity that retains its spatiotemporal unity through changes and endures through changes. It is the bearer of properties, for example, phenomenal states, memories, emotions, personal characteristics, and it possesses causal powers and is capable of making changes in the world. In my view, Strawson offers an explanationist argument for his substantivalist view on the self. According to him (Strawson, 1959, chapter three), the fact that persons are substances furnishes the best explanation for the fact that physical properties (age, height, etc.) and phenomenal states could be ascribed to the very same things, that is, individual human beings. Otherwise, that is, unless we conceptualise selves as particular substances, we cannot provide a comprehensible picture of the world.

I have various reasons for disagreeing with Strawson. First of all, I disagree with his substantivalism on the basis of the same reasons that Hume mentions when defying the substantivalist views of his period (see Sect. 1.6 in the first chapter). I do not think there is enough evidence to absolve the assumption that the self is an indivisible entity that retains its spatio-temporal unity through changes and endures through changes. Observation does not indicate that there is such an indivisible entity that retains its unity through changes. Nor could such evidence be produced by recent breakthroughs in neuroscience and cognitive psychology (from the perspective of this book, science provides a reliable criterion for the plausibility of ontology). To be fair to Strawson, he does not say that there is any kind of scientific or unscientific evidence to corroborate his view. Instead, he conjures a transcendental argument according to which, conceiving of the world as consisting of individual objects (as persons and physical things) is indispensable to making sense of the world. However, according to the scientific structural ontology that I advocate in this book, we do not need to rely on stagnant metaphysical intuitions to make sense of the world. We can use science as a guide to constructing a scientific ontology, and science can support a structuralist ontology of the world (of persons and physical things), without depending on individual objects as indispensable units (I will flesh out this structuralist ontology in the next chapter).

Generally, I do not think what bestows upon the substantivalist view its force is the strength of the argument that supports it, so much as the intuitive appeal of its general insights. Intuition, however, is not a reliable guide to ontology. I have already discussed this point in the previous chapter when I showed that what fundamental physics tells us about the basic features of reality is not in harmony with the habituated intuitions about the individual objects as the legitimate inhabitants of the domain of reality (see Ladyman, 1998; Ladyman & Ross, 2007). The same could hold true of the substantivalist theory of the self.

The substantivalist conception of the self is in harmony with the common methods that people use when they conceive of themselves. People naturally think that their selves, as individuals, endure through changes— one's self remains essentially the same from childhood to old age—and selves bear properties such as age, height, the colour of one's eyes, and so on, as well as the capacity for action and phenomenal states such as consciousness and certainty. Thus, the orthodox substantivalist picture is in line with folk psychology.

Let me clarify my point with a reference to the philosophy of modern physics once more. Contemporary physics has not been kind to the concept of individual objects. As I explained in the previous chapter, being optimistic about the picture that fundamental physics draws of the world, structural realists such as Ladyman and French endeavoured to dispense with the object-oriented view of entities and reconceptualise objects in structural terms. The lack of scientific support for the substantivalist conception of self may demand a similar solution. The motivating insight of this book is that the self is the infrastructure that underpins various properties and aspects of the self. Unfortunately, substantiating this claim is not as simple as its primary statement. In this chapter and the next one, we have to take pains to fetch the relevant pieces of argument from neuroscience and philosophy and assemble them to show how they support a structural realist picture of the self in the face of its alternative, that is, substantivalism. To be clear, it is relatively easy to show that the substantivalist theory of the self is not plausible, given the lack of scientific support for the existence of the substantial self. It will be comparatively harder to show that a structural realist theory of the self provides a plausible account of the self in the face of scientifically-informed alternatives such as pluralism or eliminativism—which both could prevail as soon as the substantivalist view was disconfirmed. To illustrate this difficulty, consider the following.

The self is not an observable entity. We cannot perceive the selves through perception. Even introspection does not allow us to discern the self clearly as a substance in the classical sense acclaimed by classical views. Selves are theoretical (i.e., unobservable) posits. The question is whether our theoretical descriptions of the self allow us to characterise the self in terms of the orthodox substantivalist picture. Does self, as a theoretical posit of neuroscience, need to be understood in terms of substantivalism? In the remainder of this chapter, I refer to some scientific accounts of the self (and scientifically informed philosophical accounts) to demonstrate that some of the most viable theories of modern psychology disregard the existence of the substantial self. Of course, we may insist on domesticating the scientific picture into the substantivalist point of view regardless of the fact that the scientific picture is not committed to a self-substance. However, I sympathise with a naturalistic view according to which it is not reasonable to give way to adulterating the scientific result to fit in with the stagnant metaphysical understanding of the phenomena, even when the stagnant understanding is in harmony with folk psychology. Be that as it may, in the next sections, I draw attention to some creditable scientifically informed accounts of the self that diverge considerably from the substantivalist view. For one thing, this survey demonstrates that there are indeed viable alternatives to the substantivalist view of the self. For another, it turns out there is a state of underdetermination of the ontological accounts of the selfhood with the puzzling diversity of scientific accounts of the self. Thus, a form of metaphysical underdetermination raises its head. Pluralism and eliminativism are among the most prominent alternatives to substantivalism that will be reviewed in this chapter, but since these theories lead to conflicting ontological accounts of the self, they cannot inform a unifying account of real features of the self. Under the circumstances, a realist must find a way to reconcile the inconsistencies and find a grip on a common structure which could underpin ontological disagreements of pluralism and eliminativism. I shall proceed with introducing the conflicting alternatives first.

3.2 Scientific Pluralism in Cognitive Science

In this section, I outline the pluralist alternative to the substantivalist conception of the self. In the previous section, I extolled naturalism—that is, the attitude of taking the philosophical implications of scientific theories seriously—over unshaken loyalty to stagnant metaphysical and folk psychology. I suggest that when there is an inconsistency between orthodox metaphysical views and ontological implications of scientific theories, one may try to find a way to tailor philosophical views to the result of reliable scientific findings. In the case of the theory of the self, too, when there is an apparent inconsistency between the substantivalist view and the implications of scientific theories, one may inform one's philosophical views by scientific theories. However, a serious problem raises its head when we surmise that not only science could be inconsistent with the orthodox philosophical views, but also scientific theories themselves come with diverging and at times conflicting implications. Here, the main reservation is that because various aspects of the self, as being explored in different scientific endeavours, do not contribute to providing a unified and consistent conception of the self, we cannot reach a unified ontological model of the self.

Various aspects of the self are explored by different methodologies and experimental tools and the results could be stated in different theoretical frameworks. The emerging pluralistic picture not only renounces substantivalism but also bypasses any theory that delineates the self as a unified entity. For, this approach presumes that vagaries of theoretical enterprises in the field of cognitive sciences do not systematise into a unifying theoretical venture with univocal theoretical outputs about the nature of the mind in general and the self in particular. The plethora of various theoretical accounts and experimental methodologies do not submit a unified picture of the self. There is no telling though that the plethora of various accounts can be incorporated into a unified model of the phenomenon, that is, the self. This line of thought will result in a pluralistic conception of the self.

In general, scientific pluralism is an approach to the philosophy of science that presumes that our scientific enterprises are dis-unified activities that aim to unfold diverse patterns and mechanisms by employing heterogeneous theoretical and experimental venues, methodologies, and interventions. One may feel a great urge to imagine scientific activity as an ideally unified endeavour. Nevertheless, the pluralist view presumes that far from giving way to the philosophical impulse to see unity where there is actually heterogeneity, sciences—even at the level of fundamental laws of physics—actually represent a world that consists of unpredictable patterns, discontinuities, and diversities (Cartwright, 1999). The philosophical urge for representing science as a single unifying project does not conform to what is observable in actual scientific practice. It glosses over the diversities of methods of classification and experimentation to the effect that it overlooks the heterogeneous nature of the things that constitute the world (Dupré, 1993). Evidently, pluralism comes with an antireductionist agenda. It is based on an understandable insight into the irreducible diversity of theoretical enterprises that aim to reveal facts about the world, that is, facts that might well be heterogeneous. I have alluded to the pluralist approach in cognitive science before, in Sect. 2.5 of the previous chapter. Below, I shall develop this point before adding further details to show how the pluralist view extends into the field of the philosophy of the self.

The pluralistic approach to cognitive sciences draws its force from the following observations. The theoretical models that capture various cognitive phenomena—or even the same cognitive phenomenon—are only loosely connected. These various theoretical models lead to multiple explanatory perspectives that may somehow be integrated without being incorporated into a unified theoretical framework (Dale, 2008). Integration simply draws loose connections between diverse theoretical enterprises on the basis of practical considerations, without going so far as to indicate that there are facts or structures that underpin genuine theoretical relations between heterogonous theoretical endeavours. Generally speaking, pluralist "integration"—which aims to establish the coherency of multiple aspects or explanations (Miłkowski, 2016)—does not need to undermine the pluralist cornerstone of Gallagher's pattern theory, which will be explicated in the next section. This means that we can support an account of integration without also presuming that there are factual relations between diverse theoretical streams, there will be no unifying framework to accommodate convergent epistemological and ontological views about the phenomena in question.

Pluralism in the field of cognitive sciences can be justified on the basis of "the range and complexity of cognitive science's subject matter" (Dale, 2008, p. 156). That is to say, cognitive science aims to model and explain complex and sophisticated phenomena, such as the brain, cognition, phenomenal experiences, perceptions, action, and behaviour. Inexorably, however, the scope of each of the theories that aim to deal with a part of the wide range of cognitive phenomena is limited. Methods, measures, scales, and mathematical tools that each set of these theories use are tailored to the specific interests and explanatory goals of the scientists that work in each subfield. When seen in this light, it appears that "cognitive science" is an umbrella term which covers a number of different subdisciplines—for example, evolutionary psychology, computational neuroscience, and so on—with diverse connections with other fields. For example, evolutionary psychology has its roots in theoretical biology, and theories of social cognition have links with social science and anthropology, and so on. This may indicate that theoretical diversity in the field of cognitive science is quite genuine and that the unification of cognitive sciences under the rubric of an all-inclusive psychological discipline with restrictive borders is unachievable. To be clear, the pluralist approach allows for methodological integrations—that is, combining the methods and findings in a piecemeal manner—without committing itself to a unificationist agenda. As Chemero observed:

> So computational cognitive scientists (e.g., Fodor and Pylyshyn, 1981) argue against the ecological approach (e.g., Gibson, 1979) as part of an effort to establish their approach as a unifying paradigm for the discipline, and hence to attract research funding and good graduate students. Ecological psychologists (e.g., Turvey et al., 1981) reject these arguments because they share few of the assumptions that structure the computational cognitive scientist's argument. And both sides continue with their experimental research, ideally having clarified and adjusted their own assumptions based on the critique from the other side. (Chemero, 2009, p. 16)

In short, cognitive science consists of heterogeneous theoretical enterprises none of which provides a uniquely comprehensive framework for the study of cognitive phenomena, because one theoretical framework cannot subsume the diversity of facts that must be explained by cognitive science (Chemero & Silberstein, 2008). Pluralists argue that it is even hard to reach a consensus on a unifying schematic that can subsume competing theories that aim to address the same cognitive phenomenon (or a class of similar phenomena) (Dale, Dietrich, & Chemero, 2009, p. 2).

3.3   Pluralism About the Self (Pattern Theory)

The pluralistic approach in cognitive science can be stated more specifically in terms of a pluralistic approach to understanding the concept of self. An excellent version of this specific form of pluralism is developed in Shaun Gallagher's pattern theory of self (Gallagher, 2013). Diverse theoretical enterprises in neuroscience cannot be incorporated into a unique all-inclusive framework, and the pluralist approach recognises competing theories that describe the same phenomena without insisting on drawing strong theoretical connections between them. Competing theories represent their topic as a bundle of heterogeneous events and processes that do not need to be aspects of a single substance or entity. The pattern theory applies this view to the concept of the self. The theory indicates that various aspects of the self can be (and indeed are) modelled in a number of different ways through different scientific enterprises, for example, neurology, behavioural science, semiotics, sociology, biology, cultural studies, and so on. Here, the general insight is that various scientific and philosophical ways of thinking about the self contribute to fleshing out different aspects of the self without incorporating them into a unique entity. Gallagher's cataloguing ascribes following aspects to the self:

> [T]he cognitive self, the conceptual self, the contextualized self, the core self, the dialogic self, the ecological self, the embodied self, the emergent self, the empirical self, the existential self, the extended self, the fictional self, the full-grown self, the interpersonal self, the material self, the narrative self, the philosophical self, the physical self, the private self, the representational self, the rock bottom essential self, the semiotic self, the social self, the transparent self, and the verbal self. (Gallagher, 2013, p. 1 emphasis original)

Notice that Gallagher's list includes "cognitive self", "conceptual self", and so on, instead of "*the* cognitive aspect of *the* self", "the conceptual aspect of *the* self", and so on. The very statement of the situation indicates that different aspects cannot be included in a comprehensive pattern of *the* self.

In harmony with the tenet of scientific pluralism, Gallagher argues that various aspects of the self, as being modelled through different scientific enterprises do not need to be conceptualised as contributing to the formation of a substance that has its own independent existence (Gallagher, 2013, p. 3). This view on the nature of the self is in conflict with the orthodox substantivalist view, according to which the world includes selves as substances that are the bearers of modes and attributes. Gallagher does not presume that the self is a real substance to which various self-patterns need to be related in order to exist. Unlike Strawson (1959), he does not even submit that we need to *understand* selves as real substances. He assumes that we are allowed to think of aspects of the self as being identifiable in terms of different patterns, without assuming that aspects need to be related to or depend on a self-subsistent entity in order to exist or to be recognised. Given the tenet of pluralism, we cannot presume that any of these patterns can capture the essence of self exclusively. Self-patterns do not even need to be incorporated into a unifying all-inclusive framework which can subsume various patterns of what a realist may want to call "*the* self". The pattern theory holds that we can identify various aspects of the self as relatively independent patterns that do not depend on a substance-like entity for their existence, although they may be loosely connected to one another to form a cluster which could be called the self. It is important to notice that there is no reason to identify the patterns that form the cluster that we usually call the self as aspects of a unique substance. According to Gallagher's statement of the theory:

> What we call self consists of a complex and sufficient pattern of certain contributors, none of which on their own is necessary or essential to any particular self. This is not a pattern theory of "*the* self." Rather, what we call "self" is a cluster concept which includes a sufficient number of characteristic features. Taken together, a certain pattern of characteristic features constitutes an individual self. (Gallagher, 2013, p. 3)

Gallagher's pattern theory could be understood as a meta-theory that catalogues elements of self and captures the maps of all possible self-patterns, without making any remarks on the sufficient or necessary condition of the self. Gallagher convincingly argues that this theory would help us to deal with cases that remain inexplicable by the orthodox approach, for example, the borderline cases. The substantivalist view could not deal with borderline cases such as dissociative identity disorder, where the appearances suggest that more than one self is involved in causing phenomenal states. The pattern theory could deal with the borderline cases because it presumes that what can be discerned as the individual self in a given situation is the result of dynamical interactions between various constituent aspects (or patterns), without presuming that the patterns must be aspects of one single unified entity. In this vein, Gallagher's theory provides a conceptual framework for speaking of dissociated selves. The substantivalist view fails to do so because it assumes that the self is a substance. Because the substance is conceptualised as an indivisible individual object, the substantivalist view cannot assimilate the notion of dissociated selves. The pattern theory, on the other hand, can clearly account for dissociated selves as clusters that include some self-patterns but not others. Two or more clusters could jointly embed the mental and phenomenal states of what we identify as a person in a social context. The pattern theory could also accommodate the notion of the self that is at issue in some schizophrenic symptoms (when the sense of agency is lost) or in some neurodegenerative diseases (that include personality change). Because the substances are indivisible units, substantivalist views cannot account for the existence of selves with lost essential aspects. Assuming that agency or memories are essential aspects of the self, the self-substance cannot lose them without disintegrating altogether. However, according to Gallagher's view, losing some aspects of self, for example, the feeling of the agency or one's memories does not mean that all is lost. The dynamical relation between the involved patterns—for example, minimal embodied aspects, experiential aspects, affective aspects, intersubjective aspects, extended aspects, and so on—may result in the formation of different kinds of self-patterns, none of which is contributing to the generation of a substantial entity. It is worth mentioning that, Gallagher's pluralistic approach to the self allows for the weak integration of self-patterns. Gallagher asserts, after all, that various self-patterns are related together to form a cluster that we may understand as the self. The weak form of integration that could take place here is piecemeal and local. Integration cannot be pursued so far as to attach self-patterns to a unifying framework or make them contribute to the formation of a substantial self. Therefore, while pattern theory aims "to show how various aspects of self may be related across certain dimensions" (Gallagher, 2013, p. 1), it does not develop this quest into a full-fledged unifying framework for embedding the individual self whose aspects are connected together strongly. To make a long story short, the pattern theory assumes that diverse self-patterns do not provide a substantial self. Nor does it support any other unifying account of the self.

As I have remarked before, scientific pluralism allows for some piecemeal forms of integration too (also see Mitchell, 2012; Mitchell & Dietrich, 2006). However, accounting for relations between various patterns, for example, in terms of a dynamic systems theory, does not ground the strong form of unification that is required for representing the self as a unified and indivisible substance that has its own independent existence. Thus, despite making room for some forms of integration, Gallagher remains loyal to the goal of pluralism. As he remarks at the end of his paper, "The important move here is to admit that there are multiple processes that may count as self-related, even if not self-specific, and that they can be constitutive of self over and above the first-person perspective. That sends us back to a pluralist approach [...]" (Gallagher, 2013, p. 5). It is also worth mentioning that in a recent paper, Gallagher (Gallagher & Daly, 2018) has elaborated on three points that allow for tracing the dynamical relations in a self-pattern (and thereby consolidating the integrative core of Gallagher's original theory). The three ways which can be used to consolidate the integrative power of self-patterns consist in considering the power of narrative, the study of psychopathology, and predictive processing mechanisms in the brain (for details see Gallagher & Daly, 2018). I will speak about the relationship between the brain's predictive processing capacity and the self in the next chapters of this book. For the time being, though, suffice it to say that Gallagher and Daly's attempt at reinforcing the integrative impact of the pattern theory does not go so far as to provide a clear account of the relationship between various self-patterns. This means that the improved version does not encroach upon the pluralist tendency of Gallagher's original approach. In the next section, I will review an alternative to the pluralist theory of the self.

3.4   Eliminativism About the Self (Being No One)

In the previous section, I referred to Gallagher's scientifically informed speculation about the self. As his theory clearly indicates, scientific research does not support the substantivalist conception of the self. It does not exonerate the philosophical view about the existence of a self-substance along the lines that are delineated in Aristotelian or Cartesian philosophy. One may react to this situation—as Gallagher did—by saying that the self is an amalgam of diverse patterns whose characteristics are explored through various streams of scientific research, without indicating that there is a unifying framework that could incorporate various patterns into a single conception of the self. Alternatively, one may take an eliminativist approach and deny that there exist such things as selves. Thomas Metzinger has advocated such an eliminativist theory of the self, and he has suggested that it is possible to renounce the substantivalist conception of the self completely (Metzinger, 2003, 2009).

The rise of cognitive science has always been associated with an eliminativist attitude towards entities that had been taken for granted by habituated intuitions and folk psychology. Perhaps a best philosophical exemplar of this attitude could be found in the philosophy of W. V. O. Quine. Influenced by the rise of behaviourism and its thoroughgoing loyalty to the goal of scientific objectivity, Quine (1960) strongly supported dispensation with mental objects and mental states such as beliefs, which could not be identified with the physical states or behavioural patterns. According to this view, the mentalist discourse has to be renounced in favour of a scientific discourse, which is based on a vocabulary of verbal behaviours or neurophysiological states as suggested by the best scientific theories of our (i.e., Quine's) time. This last assertion is derived from a centrepiece of Quinean naturalism, according to which philosophy can be fertilised by the application of the scientific methodology and our best scientific theories tell about the world. When applied to the field of philosophy of mind, this naturalistic attitude indicates that philosophy must draw on the best available scientific theories to replace stale ways of speaking about mental objects and phenomenal properties with the accurate characterisation of the underpinning neurological mechanisms. Of course, there are a number of different ways for fertilising philosophy of mind by recent breakthroughs in cognitive science. It may be possible to invoke bridge laws to translate the vocabulary of the old theory in terms of the new vocabulary the scientific research generates. It is possible to use different strategies to reduce the statements of the old theory to laws and theories of cognitive science. This latter approach is tagged reductionism, but the attitude that the eliminativist exercises is a bit more radical.

According to eliminativism, for the goals of a full-blooded version of naturalism to be achieved, appealing to reductionism is not quite enough. The terms in the vocabulary of folk psychology cannot be simply identified with or reduced to brain states or underpinning neurological mechanisms. This is because folk psychology is too poor to afford theoretical terms that can be reduced to or identified with theoretical statements of modern psychology. If traditional psychology had accommodated relatively viable theories, it would have been possible to reduce traditional psychology to what is revealed by recent cognitive neuroscience. However, given the theoretical poverty of folk psychology, there is no prospect of a match (or intertheoretical reduction) between folk psychology concepts and theories from recent cognitive science. Eliminativism is even more radical than reductionism. Barring the possibility of reduction, the eliminativist argues that mental states and mental properties, which are recognised as being imprecise and at times misleading theoretical terms, must be eliminated altogether. A similar eliminativist approach has been advocated by Paul Churchland, who suggests that the discourse of folk psychology—for example, references to propositional attitudes—does not produce a valid theoretical vocabulary, and thus its terms and statements cannot be used in plausible psychological explanations (Churchland, 1981). This is because, as Churchland observes, our best scientific theories (in computational neuroscience and neurology) about the operation of the brain and nervous system do not support the folk psychological conception of thinking in terms of processing propositions or beliefs. Cognitive neuroscience does not justify the identification of thinking mechanisms in terms of the functioning of syntactical engines which process language-like entities. Nor does it support the sentential and doxastic accounts of epistemology that misleadingly suggest that thinking consists of the processing of language-like entities (i.e., propositions) or beliefs (Churchland, 1981, 1989). Folk psychology is not only an incomplete source of information about cognition and mindfulness, but it also provides a misleading and confusing description of the mind and its states. In all, the eliminativist approach presumes that parts of the theoretical vocabulary of the stagnant psychology that cannot be converted to theories of the cognitive sciences must be removed from the intellectual discourse, in the same way, that concepts such as "phlogiston", "vital spirits", and "demonic possession" are eliminated. Concepts such as phlogiston and vital spirit proved to be inconsistent with the vocabulary of advanced, mature scientific theories and thus have been removed from the intellectual discourse (Churchland, 1981). The same should happen to the concepts of folk psychology. Now, let us assume that the "self" is a concept of folk psychology that cannot be relocated in the theoretical vocabulary of the recent cognitive neuroscience. That is to say, there is no counterpart for the notion of "self" in the vocabulary of mature and advanced theories of modern psychology. On such grounds, one may produce a plea for eliminating the "self" from the intellectual discourse.

It is possible to diverge from the substantivalist view in a number of different ways. Thomas Metzinger took a bluntly eliminativist approach to the substantivalist conception of the self (Metzinger, 2003, 2009). Metzinger's account is eliminativist about the self in the same sense that Quine's and Churchland's views are eliminativist about beliefs and mental states. Metzinger's theory assumes there is no one-to-one match between the substantivalist notion of the self and the underpinning neurological processes that are revealed by respective scientific theories. Our neurological explorations of the brain do not reveal anything resembling a substantial self. It follows that, to the extent that the results of scientific explorations are concerned, there exist no such things as selves in the world. The substantivalist self is not a valid theoretical entity, and it could be dispensed with in either scientific or philosophical contexts (Metzinger, 2003).

After eliminating the self from his ontology, Metzinger endeavours to account for the self's experiential states in terms of sufficient and necessary conditions that contribute to the formation of self-models.

Self-models are models whose representational content cannot be recognised as representational by themselves. Phenomenal self-models are a special kind of self-model. A phenomenal self-model "internally and continuously simulates its own observable output as well as it *emulates* abstract properties of its own internal information processing—and it does so *for* itself" (Metzinger, 2003, p. 301). The most important property of phenomenal self-models is their transparency. A phenomenal self-model's transparency is due to "the *attentional unavailability of earlier processing* stages for introspection" (Metzinger, 2003, p. 165 original emphasis). Phenomenal self-models are computational modules (or integrating processes) capable of embedding intentional relations and selective attention (Metzinger, 2003, chapters 3 and 6) (also see Blanke & Metzinger, 2009; Lenggenhager, Tadi, Metzinger, & Blanke, 2007). The biological system's capacity for having internal temporal representations (or "the windows of presence") is among the multiple constraints of having a phenomenal self-model. The same holds true for the system's capacity for integrating its internal representations with the world-model that makes the current situation globally available to the system. The situation must be globally available to the system so that the system can render information processing, which realises cognitive references (to the world) or controlled actions (on the world) (Metzinger, 2003, p. 119). To interact with the world, the system activates its self-model, that is, a transient computational module, which runs sensorimotor mechanisms and makes selective attention possible. In this fashion, Metzinger endeavours to account for the functional and representational properties of the self without recognising the "self" as a legitimate theoretical term in the scientific discourse or presuming that the term refers to an actual entity in the world. The self-model can be used efficiently enough for theoretical purposes, such as accounting for the representational states of the self—which can be individuated on the basis of their content—and functional states—which are individuated on the basis of their causal roles. Self-models can also be used for practical purposes, for example, diagnosing neurological disorders (heautoscopy or out-of-body experience) (Blanke & Metzinger, 2009). Even sophisticated phenomena such as the emergence of societies can be explained by using self-models. That is to say, Metzinger's theory relies on self-models not only to account for cognition, perception, and behaviour of individual persons but also to deal with the question of foundations of social relations. The theory presumes that the self-model's capacity for simulating another self-model's internal representations is enough to account for the constitution of the web of social relations (Metzinger, 2003, p. 367 ff.). Metzinger's account is nicely adorned by neurological details about the operation of the underpinning neurobiological mechanisms that realise different aspects of the self-model. For example, he argues that the self's simulating capacity is based on the working of the Mirror Neuron System (Metzinger, 2003, p. 386). According to this theory, the Mirror Neuron System is supposed to be a part of unconscious self-model, which can be used to model the internal models of another self-model as a part of the external world or social reality (ibid.).

Metzinger's theory is admirable on account of its reliance on advanced neurological theories of the field, as well as its loyalty to the spirit of scientific philosophy. That being said, I also have to point out that his theory suffers from some minor hiccups, which could result in serious disfluencies in providing a convincing account of the self-model. In the next two sections, I will state some reservations as regards the completeness and plausibility of the pattern theory and the eliminativist theory of the self. I also express some reservations about the ontological consequences of the inconsistency between these theories. To be more precise, I shall argue that the divergence of these two theories could result in a state of metaphysical underdetermination of the self. In the next chapter, I will address this problem and argue that the state of underdetermination can motivate a structural realist theory of the self.

3.5   Evaluating the Eliminativist View

Metzinger denies that there are such things as selves in the world. However, he does not go so far as to deny that there are representational, functional, and phenomenal properties that could be assigned to the so-called self. The situation gives rise to a question that is acknowledged by Metzinger himself. As Metzinger acknowledges, "the pivotal question is, what justifies treating all these highly diverse kinds of information and phenomenal representational content as belonging to *one* entity?" (Metzinger, 2003, p. 302). Metzinger endeavours to address this question by appealing to the property of "mineness". But mineness is itself a relational property. It indicates that something belongs to (my)self. Therefore, it should be understood as a relation (i.e., the possession relation) between the self (or myself) and something else. As Metzinger ingenuously improvised, the property of mineness can be used as a cohesive glue that puts together all cognitive information processing mechanisms, mental content and phenomenal properties. However, it does so by showing that all of these mechanisms and properties are related to the self (myself). Thus, it takes the existence of that something—that is, the self and perhaps even a substantial self—for granted. This poses a scholastic threat to Metzinger's attempt for eliminating the self; the property of mineness as a relation takes the existence of the relatum (the self) for granted. I have little sympathy with scholastic scruples of this kind. However, I have to submit that Metzinger's philosophical arguments are not developed to ward off this objection, and his employment of the property of mineness by itself (i.e., without being related to a self) does not provide enough grounds for accounting for diverse aspects of the so-called self. It is true that Metzinger endeavours to find a placeholder for the self. He argues that what is commonly called the self is not a substantial self but a self-model. After contriving the self-model, it would be possible to account for the mineness in relation to the self-model, instead of a substantial self. The self-model is not the substantial self and according to Metzinger, there are no such things as selves. As I remarked, he suggests that the self-model could take the place of the self for all philosophical and scientific purposes. However, if it is indeed the case that the self-model can serve as the self for all scientific and philosophical purposes, why not go further and identify the self-model with the self in all scientific and philosophical contexts? That is to say, why eliminate the substantial self from the ontological sphere altogether, instead of reducing it to the so-called substantial self or identifying the two with one another? If two things, or rather two terms referring to two things are interchangeable in all scientific and philosophical contexts, surely we do not need to eliminate one of them and put the other in its place. We can simply presume that two entities that are referred to by two interchangeable terms are identical. Let me elaborate.

One central claim of the eliminativist approach is that the theoretic vocabulary of the old theory can't be reduced to or identified with the components of the new theory. The central theoretical terms of an expired and false scientific theory—for example, "phlogiston" in the old theory of combustion—cannot be reduced to theoretical terms of a new theory— for example, reduction-oxidation reaction. In view of the impossibility of intertheoretical reductions and identification, and owing to inappropriateness of the old terms, the eliminativists suggest that some central theoretical concepts of the old theory must be disposed of. The question is, does this situation apply to the case of the self (and its elimination)? In my opinion, the case of the self does not warrant the application of eliminativism. It is true that the self-model is not precisely the same thing as the good old substantial self. However, the situation is not dire enough to warrant the elimination of the self altogether. Even if we concede that the scientific image of the self is not in line with the substantivalist conception of the self (as I concede), it will still be possible to presume that the substantivalist conception could be reduced to a scientifically respectable conception of the self. Reductionism respects the difference between old and new theoretical concepts. However, even after considering the divergence between old and new theories, it would still be possible to reduce the elements of the old theory to the vocabulary of the new theory, instead of eliminating the theoretical terms of the old theory altogether. My point is that, as Metzinger's own statement shows, the concept of the self is indeed reducible to the new notion of the self-model and could be identified with it, precisely because it has been assumed that for all scientific and philosophical purposes, the self could be replaced with (i.e., identified with, reduced to) the self-model. It follows that it is possible to reconceptualise the classical self in terms of the self-model, which is essentially different from the substantial self, without going so far as to eliminate the self-substance altogether. Thus, in a nutshell, Metzinger's account does not eliminate the substantial self so much as reconceptualises it on the basis of information provided by neurology and computational neuroscience.

Now let us return to the question that has been stated at the beginning of this section. The question was: "What justifies treating all these highly diverse kinds of information and phenomenal representational content as belonging to *one* entity?" Metzinger's reply—which mentions the property of "mineness"—does not take the role of the self into account. It does not presume the self is the bearer of kinds of self-related representational content and functional properties, and it does not presume that it is in virtue of the self that we can justify our ascription of representations and information to one entity. Metzinger dismisses this possibility because he is an eliminativist about the self. However, in light of the argument unfolded in the previous paragraph, it can be observed that Metzinger's approach does not warrant the elimination of the self so much as paves the way for its reduction or reconceptualisation in terms of the self-model. So, to the question that was stated at the beginning of this section, we can still reply by referring to the fundamental role of the self. However, we have to be clear about the difference of the new scientifically informed concept of the self—to which we reduce the old concept—and the substantivalist conception of the same entity—which must be reduced to the new concept. In short, the eliminativist theory cannot connect different streams of self-related information processing, representational content, and functional structures to each other without invoking a self-model. Nevertheless, as the self-model identifies with the self—for all philosophical and scientific purposes—it can be contended that Metzinger smuggles in the self-concept through the back door. The eliminativist theory does its best to inform the philosophical conception of the self by recent scientific breakthroughs. However, it fails to deliver on its promise of eliminating the self from science and philosophy.

3.6   Evaluating the Pattern Theory

The pattern theory comes with the expressed aim of interpreting different aspects of the self as compatible and commensurable, instead of opposing or inconsistent. There are numerous processes and mechanisms that may count as self-related, and one virtue of Gallagher's pattern theory is that it endeavours to draw meaningful connections between these aspects, without indicating that they belong to an independently subsisting entity. However, more work is needed to show how the theory draws meaningful connections between aspects of the self without presuming the existence of the substantial self. That is to say, without assuming that there is a substantial self that subsumes various aspects of the self and bears them, various aspects of the self cannot be related to each other in a meaningful and philosophically interesting way. Notice that the substantivalist view could account for the unity of the self and the relationship between its various aspects by holding that the aspects are related to the self-substance and are borne by it. Now, if we remove the self-substance from the picture, the self gets disintegrated. This would pose a threat to Gallagher's claim as to the existence of meaningful relations between various aspects and features that constitute what can be understood as an individual self. Thus, in spite of paying lip service to the idea of setting down meaningful relations between different aspects of the self, Gallagher doesn't tell a full story about how various elements in the self-patterns (or rather various self-patterns themselves) are connected to form an individual person (Beni, 2016, p. 3731).

As I remarked before, more recently Gallagher (Gallagher & Daly, 2018) went out of his way to address this criticism. He has reacted to it by suggesting that there are various ways of accounting for the dynamical relationship between self-patterns. It is indeed a sign of Gallagher's broadmindedness that he has recognised the problem. However, in his recent enterprise (Gallagher & Daly, 2018), he merely emphasises the fact that there are relations between self-patterns, instead of actually providing a viable and detailed account of relations between various patterns within the context of his pattern theory. Gallagher invokes three strategies to establish the points that dynamical patterns must be somehow related. For example, he suggests that the narrative aspect of the self, "offers a way to map the dynamical relations among the various other factors of the self" (Gallagher & Daly, 2018, p. 5). However, as Gallagher's original statement of the pattern theory in his 2013 paper indicates, the narrative self is just one pattern theory among numerous others, and thus it cannot subsume other aspects (or patterns) of self and connect them together. A given self-pattern (such as the narrative self-pattern) cannot take a privileged status to govern the interrelations between other aspects and patterns of the self. It is possible to evaluate in the same way the two other strategies that Gallagher proposes to account for the relation between self-patterns. For example, Gallagher and Daly suggest that predictive coding theory could help us to specify dynamical relations that constitute self-patterns (Gallagher & Daly, 2018, p. 1), and I am in complete agreement with this suggestion. However, their suggestion is not spelt out expansively enough to show how predictive processing theory (as a neuroscientific theory) could account for the relation between various self-patterns. The neuronal self-pattern that is fleshed out on the basis of neuroscientific data is just one pattern among many, and it does not need to subsume all other self-patterns and account for their interrelationship. Thus, Gallagher and Daly do not show how it is that this specific account of the nature of self-patterns (in terms of predictive processing) fits in with the central tenet of the pattern theory. Under the circumstances, the integrative strategies that Gallagher has recently suggested do not succeed at overcoming the pluralistic core of his theory. I will elaborate on my evaluation of Gallagher and Daly's paper in Sect. 4.3 of the next chapter.

My opinion is that in order to be able to account for the relationship between different self-patterns in a meaningful and philosophically interesting way, we must concede the existence of some kind of central self-concept. The problem is how to characterise this central self-concept without making commitments to substantivalism, given that the substantivalist conception of the self receives little support from contemporary cognitive neuroscience. This book aims to provide a clear account of the required self-concept, which is arguably different from the substantivalist self. To do so, we have to strike a fine balance between the orthodox metaphysics of the self—which presumes that there is a self-substance—and scientifically informed theories that either eliminate the self or suggest that the self is a disintegrated (or rather loosely integrated) cluster of various patterns. To do so, I suggest that it is best to reconceptualise the self in terms of a structural entity, instead of an individual object or an orthodox substance. The assumption of the existence of a central self-concept—which could be specified structurally—would help us to account for the relationship between different aspects and elements of self, without making commitments to the substantial self. I shall unfold my structural realist theory of the self in the next chapter. Before that, in the remainder of this chapter, I explain how it is that the theoretical diversity of scientific accounts of the self actually paves the way for the genesis of a structural realist theory of the self.

3.7   Some Other Theories

It is worth mentioning that the eliminativist theory and the pattern theory are only two amongst various theories that seek to account for the self and its various aspects scientifically. There are several other viable theories of self, which will be mentioned presently. Here, it is important to notice that such theoretical multiplicity increases the force of metaphysical underdetermination, which wreaks havoc with the orthodox ontology of the self (I will elaborate on underdetermination in the next section).

The CMS theory of the self, as being defended by Georg Northoff and colleagues (Northoff & Bermpohl, 2004; Qin, Duncan, & Northoff, 2013), is one of the other viable theories. The theory emphasises the role of CMS (cortical midline structures of the brain, consisting of the medial prefrontal cortex, anterior cingulate cortex, and posteromedial cortices) in the processing of pieces of self-related and self-specific information. According to this theory, the self-concept could be specified on the basis of the processing of self-related or self-specific information in various cortical midline regions of the brain, for example, the ventromedial prefrontal cortex (VMPFC), the anterior cingulate cortex (ACC), the dorsomedial prefrontal cortex (DMPFC), and the posterior cingulate cortex (PCC). Self-specific stimuli are a type of stimuli "that are experienced as strongly related to one's own person" (Northoff et al., 2006, p. 441). The CMS theory receives remarkable experimental support from brain imaging studies that draw a connection between the neuronal activity of medial cortical regions (e.g. the perigenual anterior cingulate cortex, dorsomedial prefrontal cortex, and the posterior cingulate cortex) and the subject's self-related and self-specific experiences (Beni, 2016). Some of these studies indicate that self-referential processing is functionally dissociable from other forms of semantic processing within the human brain. For example, an experiment conducted by Kelley et al. (2002) proposes to categorise the subjects' judgements about the trait adjectives under three experimental classes. The classes subsume categories of self-relevant judgements, other-relevant judgements, or case judgements. The results of Kelley et al.'s brain imaging experiment indicate the medial prefrontal cortex is specifically active during the processing of the self-related judgements. This means that CMS is active during self-specific information processing. There are similar experimental studies to the same effect. In another experimental setting, researchers scanned subjects' neural reactions to emotionally significant pictures (of oneself or those close to oneself). The experiments indicate that a patient's CMS is active during the perception of his or her own picture or self-related images (Qin et al., 2013). Other studies indicate that CMS gets stimulated in reaction to self-related stimuli even when the person is in an unconscious vegetative state (Qin et al., 2010). Further studies indicate that it is possible to draw a connection between the content of self-referring autobiographical memory and of episodic memory and information processing of CMS activation (Summerfield, Hassabis, & Maguire, 2009). On such grounds, Northoff and colleagues claim that the processing of the self-specific information in CMS underpins our conception of the self. The CMS theory provides a viable alternative to the pattern theory and the eliminativist theory of the self. And there are yet further viable theories of selfhood.

Vittorio Gallese's Embodied Relational Self (ERS) is yet another viable theory of self (Gallese, 2014). Gallese draws on resources of theories of grounded cognition and underlines the role of the body and its social context which situates the self. Gallese's theory is influenced by the embodied approach to cognition, that is, an approach which emphasises the coevolution of body and cognition. It explicates the organism-environment dynamical relationship in the context of the evolution of cognition (Niedenthal, Barsalou, Winkielman, Krauth-Gruber, & Ric, 2005; also see Pezzulo et al., 2011). The approach lays emphasis on the role of motor control and sensorimotor brain regions—for example, premotor cortex, primary motor cortex, cerebellum, and so on—in the formation of cognition, and eventually, in the formation of the self-concept. Gallese's theory of the self also relies on the situatedness of the self in the web of social relations. As he remarks, "the notion of intersubjectivity is intrinsically related to the notion of the self" (Gallese, 2014, p. 1). ERS holds that focusing on the function and the structure of the experiences of the self from the first-person perspective is barely enough to provide a reliable understanding of the nature of the self. Hence there is a need for going beyond symbolic information processing in the brain of a person and expanding the theory of the self in a way that involves the role of social and embodied factors. Accordingly, ERS embarks on specifying the self in relation to the webs of social interrelations that are associated with the operations of the bodily mechanisms. It underscores the relation of the physical self to its environment and social network. Gallese's view on the social aspects of the self is accompanied by a sophisticated account of the functioning of the Mirror Neuron System (Gallese, Eagle, & Migone, 2007).

CMS and ERS theories are not the only additional options. There are yet further theories of self that need to be surveyed in the next chapters. The predictive processing theory too can be conjured to account for the self and some of its phenomenal aspects, for example, consciousness (Apps & Tsakiris, 2014; Friston, 2018). I will discuss the predictive processing theory in Chaps. 4 and 5 of the book and will consider ERS expansively in Chap. 6. For the time being, it should suffice to pinpoint the fact that eliminativism and pluralism are only two options amongst the plethora of scientific or scientifically informed theories of the self. The increasing diversity of viable options for conceiving of the self on the basis of theories of cognitive science enhances the force of the underdetermination problem, which will be unfolded immediately.

3.8 Metaphysical Underdetermination Raises Its Head Again

In the previous sections, I have explained that both the pattern theory and the eliminativist theory explicate their divergence from the substantivalist view openly. Scientific studies that inform these theories do not support the substantivalist view of the self, and thus the substantivalist view is dispensable. However, neither the pattern theory nor the eliminativist theory could provide a viable alternative to the substantivalist view. After dispensing with the substantivalist view, the pattern theory cannot provide a viable account of the relationship between various self-patterns. Therefore, it would conceive of various aspects of the self as disintegrated or at best loosely integrated. Alternatively, unless further elaborations can be offered, Metzinger's theory sneaks the self-concept through the back-door when it comes to accounting for various aspects of the self-model. This means that none of the divergences from the orthodox substantival-ist conception of the self provide a viable alternative to it. We are in need of a well-posed metaphysical alternative to substantivalism to account for the dynamical relationship between different aspects of the self. The problem, though, is that there are various alternatives to substantivalism, and these alternatives come with conflicting philosophical implications. Let me elaborate.

In Sect. 2.2.2 in the second chapter of this book, I explained that the theoretical diversity of various formulations of many particle physics leads to the state of underdetermination of metaphysics by physics. Here, I argue that the theoretical diversity of cognitive theories of the self leads to a situation that bears an uncanny resemblance to the state of underde-termination in modern physics. For, as my review of the scientifically informed theories of the self in this chapter indicates, we face a state of the underdetermination of metaphysics of the self by scientific theories that are not consistent with one another.

In Sects. 3.2 and 3.3 of this chapter, I canvassed two scientifically informed theories of the self. These are pattern theory and self-model. On which theory shall we rely so as to address the question of the existence and reality of the self? Metzinger's eliminativism and Gallagher's plural-ism point in different directions. One of them presumes that there are no such things as selves, whereas the other presumes that an individual self is a cluster of loosely integrated numerous self-patterns. Which one of these theories provides a reliable guide to the construction of the ontological conception of the self? Please note that a scientific realist about the self would aim to base her philosophical conception of the self on what our best scientific theories reveal about the subject. Now, as far as theories of Gallagher and Metzinger are indicating, it seems that the philosophical implications of scientific theories of the self are inconsistent or even con-flicting. The realist about the self has to face a puzzling quandary; do multiple selves (as loosely integrated self-patterns) exist in the real world, or does the world include no self at all but only self-models? Ontological commitments of these theories are inconsistent or even contradictory. Also, note that an instrumentalist does not need to be worried about the situation, as from their point of view, theories are only empirically adequate tools that help us to succeed in our interaction with the world, without aiming to explain the world or describe its real features. The pattern theory and the eliminativist theory could be both adequate instruments for, say, accounting for neurological disorders or the person-world interaction, regardless of the consistency of their philosophical consequences with one another. This is alright from the instrumentalist perspective because the instrumentalist does not need to be concerned about the consistency of their philosophical consequences. However, for a realist, who assumes that theoretical posits refer to the states of affairs in the real world, the inconsistency between Gallagher's and Metzinger's respective theories could be disturbing. From the perspective of orthodox metaphysical intuitions, the ontological consequences of both theories could not be embraced simultaneously; the self could not be many things and nothing at the same time. And the scientific evidence is not enough for ruling out one of these conceptions of the selfhood in favour of the other. Both theories could account for scientific evidence about the self, if not precisely in the same way, at least with almost equal adequacy. We cannot appeal to some piece of scientific evidence to show that one of these theories has an edge over the other. The rivalry cannot be concluded on the basis of scientific data. This means that the divergence between Gallagher's and Metzinger's respective theories could give rise to an instance of metaphysical underdetermination. It is the underdetermination of the metaphysical conception of the self by various scientific theories of cognitive neuroscience.

In Chap. 2 (Sect. 2.2.2), I explained that sub-particle physics underlies diverse (and at times inconsistent) theoretical commitments to the existence (or non-existence) of individual objects. Modern physics harbours inconsistency between the individualistic and the non-individualistic conceptions of sub-particle objects (see French, 2011, 2014). Which package (individualist or non-individualistic) could inform the ontology of sub-particle physics? As French and Redhead (1988) have argued, it does not seem that scientific evidence rules out one package in favour of the other. Under the circumstances, it would not be possible to develop a consistent ontology of sub-particle physics within the framework of object-oriented or substantivalist metaphysics. This is because substantivalism does not admit of objects that are both individual and non-individual at the same time. It does not even make sense to speak of such objects in the substantivalist framework. The solution is to find a framework that could comprise ontological commitments of the individualist and non-individualistic packages. Structural realists such as French and Ladyman have suggested that structural realism provides such a new metaphysical framework.

In the previous chapter, I reviewed recent breakthroughs in the philosophy of science to show how structural realism (SR) can provide a successful strategy for facing the threat of metaphysical underdetermination in the field of the philosophy of physics. In a nutshell, the solution consists of showing that there are commonalities or common structures that underpin both individualistic and non-individualistic conceptions of objects. According to Ontic version of SR, ontological commitments have to be made on the basis of those common structures. There is a resemblance between the state of underdetermination in physics and the state of underdetermination in the neuroscientific accounts of the self. In this section, I highlighted the divergence between theoretical commitments of Gallagher's and Metzinger's respective theories, and I argued that in the field of the philosophy of self, too, a form of metaphysical underdetermination breaks out. The ontological commitments that could be associated with each one of these theories are the opposite of the commitments of the other theory. The ontology of self would include both multiple selves (i.e., pluralism)[1] and no self at all (a la eliminativism) at the same time. Putting these two theories together, it follows that different aspects of selves exist (pluralistically) and selves do not exist at all. And there are other theories of the self (e.g., CMS, ERS) which increase the diversity. The evident theoretical diversity poses a stronger threat to

---

[1] Perhaps it is worth mentioning that the issue of metaphysical underdetermination of the self could be stated only within the context of the pattern theory. That is to say, the pattern theory itself, and regardless of its inconsistency with the eliminativist account, could establish the state of theoretical diversity which lies at the foundation of metaphysical underdetermination. This is because the pattern theory presumes that the self is a cluster of multiple patterns none of which could occupy a central or privileged status to be the exclusively correct self-pattern. In this picture, the unity of the self is underdetermined by the multiplicity of various self-patterns. Here too, a structural realist could argue that the self is the common structure that underpins various patterns (see the next chapter).

the realist conception of the self, and by the same token, it motivates the search for the unity underneath the diversities. The structural realist theory of the self offers to address the issue of underdetermination. It provides a unifying account of the interrelation between various aspects of the self.

Thus the state of metaphysical underdetermination breaks out. In the next chapter of the book, I will argue that a structural realist strategy will serve our endeavour for addressing the metaphysical underdetermination of the self too.

3.9 Towards a Structural Realist Theory

Although scientifically informed theories of the self, in the works of Gallagher, Metzinger, and others, bypass the orthodox substantivalist theory of the self, they do not provide a philosophically viable alternative to it. It seems that Gallagher's theory has some problems in accounting for the relationship between various self-patterns. Similarly, Metzinger's theory must elaborate on how to account from various aspects of the self-model without letting in the substantial self from the backdoor. Also, the philosophical implications of these theories (also theories of Northoff and Gallese) are diverging, and more work needs to be done before a scientific realist (about the self) could give structure to the diverging philosophical implications of these theories and incorporate them into a viable alternative to the orthodox substantivalist theory of the self.

A scientifically informed realist theory of the self aspires to build its ontology of the self upon the best scientific theories of the field. Diverse theories, for example, pattern theory, self-model, the CMS theory, and ERS delineate what we call the self in a number of different ways. Sometimes the theoretical implications of these theories are inconsistent—this is most apparent in the case of pluralism vs. eliminativism. It might be contended that diverse pictures of the self that are depicted by these various theories cannot be all correct at the same time, and the self cannot be many patterns and non-existent simultaneously. It might be assumed that the self cannot have its seat in CMS and inhabit the sensorimotor and Mirror Neuron System at the same time. We cannot rely on all of these theories to form an educated opinion about the essence and properties of the self as a real entity. Science itself does not provide a warrant for assuming that the divergent aspects that are described by various theories are all related to one central substantial entity. Neither can we rely on decisive scientific evidence to rule out some of these theories and confirm just one of them. For, pieces of evidence seem to support diverse theoretical approaches to the self more or less with equal force. The antirealist may find a reason to revel in the metaphysical underdetermination caused by the multiplicity of viable theories. This is because antirealists generally assume that theories are just useful tools that make scientific progress possible, without assuming that theories aim to represent the features of reality or explain the self. Thus they can easily concede that, given the multiplicity and inconsistency of the theories, there is no real self to be the selfsame target of various scientific models at any rate. But the realist, who is inclined to think of theories as representing the features of reality would see a real challenge in the underdetermination problem.

As I have argued in the previous section, SR is a viable theory that can deal with the problem of metaphysical underdetermination efficiently. What must be done in the face of the underdetermination challenge, according to the structural realist strategy, is to draw on commonalities between various scientific accounts of the self so as to establish ontological commitments that are not inconsistent or contradictory. None of the theories that are mentioned in this chapter—for example, the pattern theory, the eliminativist theory, and so on—depicts by itself the self comprehensively. Each of these theories models just a part of various aspects and elements of the self. However, the realist must be able to show that various theories of the self are modelling aspects and elements of a selfsame entity. How could we construct this unifying account? The structural realist approach endeavours to produce a reply to this question, based on underlining the role of commonalities between different theories. Here, the main claim is that there is one common pattern that can be discerned in various scientific accounts of the self. According to this proposal, the realist's best chance for a viable ontological account of the self consists of using this common structure as the unique subject of the ontological commitments. As I will argue in the next chapters of this book, the structural realist strategy also helps us to account for the relationship between various self-patterns more convincingly than Gallagher's account of integration could. While the structural realist model of the self denounces pluralism, it does not aim to revert to the substantivalist picture. In the same vein, I will argue that the structural realist theory of the self could be more complete and compelling than either eliminativism or pluralism. What the structural realist requires for accomplishing her goal is to find the common underlying structure, which could be the subject of the ontological commitments. To sate the situation dramatically (by borrowing words of J. R. R. Tolkien), in order to make our ontological commitments in the face of the diversity of theoretical accounts of the self, we need to find one self-structure to underpin them all, one structure to unify them, one structure to bring them all, and in the spirit of realism bind them. In the next chapter, I will present my structural realist account of the self.

References

Apps, M. A. J., & Tsakiris, M. (2014). The Free-Energy Self: A Predictive Coding Account of Self-Recognition. *Neuroscience & Biobehavioral Reviews, 41*, 85–97. https://doi.org/10.1016/J.NEUBIOREV.2013.01.029

Beni, M. D. (2016). Structural Realist Account of the Self. *Synthese, 193*(12), 3727–3740. https://doi.org/10.1007/s11229-016-1098-9

Blanke, O., & Metzinger, T. (2009). Full-Body Illusions and Minimal Phenomenal Selfhood. *Trends in Cognitive Sciences, 13*(1), 7–13. https://doi.org/10.1016/J.TICS.2008.10.003

Cartwright, N. (1983). *How the Laws of Physics Lie*. Oxford University Press. https://doi.org/10.1093/0198247044.001.0001

Cartwright, N. (1999). *The Dappled World: A Study of the Boundaries of Science*. Cambridge University Press. Retrieved from https://www.cambridge.org/ie/academic/subjects/philosophy/philosophy-science/dappled-world-study-boundaries-science?format=PB&isbn=9780521644112

Chemero, A. (2009). *Radical Embodied Cognitive Science*. Cambridge, MA: MIT Press.

Chemero, A., & Silberstein, M. (2008). After the Philosophy of Mind: Replacing Scholasticism with Science*. *Philosophy of Science, 75*(1), 1–27. https://doi.org/10.1086/587820

Churchland, P. M. (1981). Eliminative Materialism and the Propositional Attitudes. *The Journal of Philosophy, 78*(2), 67. https://doi.org/10.2307/2025900

Churchland, P. M. (1989). On the Nature of Theories: A Neurocomputational Perspective. In C. W. Savage (Ed.), *Minnesota Studies in the Philosophy of Science, Volume 14. Scientific Theories* (pp. 59–101). Minneapolis, MN: University of Minnesota Press.

Dale, R. (2008). The Possibility of a Pluralist Cognitive Science. *Journal of Experimental & Theoretical Artificial Intelligence, 20*(3), 155–179. https://doi.org/10.1080/09528130802319078

Dale, R., Dietrich, E., & Chemero, A. (2009). Explanatory Pluralism in Cognitive Science. *Cognitive Science.* https://doi.org/10.1111/j.1551-6709.2009.01042.x

Dupré, J. (1993). *The Disorder of Things: Metaphysical Foundations of the Disunity of Science.* Cambridge, MA: Harvard University Press.

French, S. (2011). Metaphysical Underdetermination: Why Worry? *Synthese, 180*(2), 205–221. https://doi.org/10.1007/s11229-009-9598-5

French, S. (2014). *The Structure of the World: Metaphysics and Representation.* Oxford: Oxford University Press. https://doi.org/10.1093/acprof:oso/9780199684847.001.0001

French, S., & Redhead, M. (1988). Quantum Physics and the Identity of Indiscernibles. *The British Journal for the Philosophy of Science, 39*(2), 233–246. https://doi.org/10.1093/bjps/39.2.233

Friston, K. J. (2018). Am I Self-Conscious? *Frontiers in Psychology, 9*, 579. https://doi.org/10.3389/FPSYG.2018.00579

Gallagher, S. (2013). A Pattern Theory of Self. *Frontiers in Human Neuroscience, 7*, 443. https://doi.org/10.3389/fnhum.2013.00443

Gallagher, S., & Daly, A. (2018). Dynamical Relations in the Self-Pattern. *Frontiers in Psychology, 9*, 664. https://doi.org/10.3389/fpsyg.2018.00664

Gallese, V. (2014). Bodily Selves in Relation: Embodied Simulation as Second-Person Perspective on Intersubjectivity. *Philosophical Transactions of the Royal Society of London B: Biological Sciences, 369*(1644). Retrieved from http://rstb.royalsocietypublishing.org/content/369/1644/20130177.short

Gallese, V., Eagle, M. N., & Migone, P. (2007). Intentional Attunement: Mirror Neurons and the Neural Underpinnings of Interpersonal Relations. *Journal of the American Psychoanalytic Association, 55*(1), 131–175. https://doi.org/10.1177/00030651070550010601

Gibson, James J. 1979. *The Ecological Approach to Visual Perception.* Boston: Houghton Mifflin.

Kelley, W. M., Macrae, C. N., Wyland, C. L., Caglar, S., Inati, S., & Heatherton, T. F. (2002). Finding the Self? An Event-Related fMRI Study. *Journal of Cognitive Neuroscience,* *14*(5), 785–794. https://doi.org/10.1162/08989290260138672

Ladyman, J. (1998). What Is Structural Realism? *Studies in History and Philosophy of Science Part A,* *29*(3), 409–424. https://doi.org/10.1016/S0039-3681(98)80129-5

Ladyman, J., & Ross, D. (2007). *Every Thing Must Go.* Oxford: Oxford University Press. https://doi.org/10.1093/acprof:oso/9780199276196.001.0001

Lenggenhager, B., Tadi, T., Metzinger, T., & Blanke, O. (2007). Video Ergo Sum: Manipulating Bodily Self-Consciousness. *Science,* *317*(5841), 1096–1099. https://doi.org/10.1126/science.1143439

Metzinger, T. (2003). *Being No One: The Self-Model Theory of Subjectivity.* Cambridge, MA: MIT Press.

Metzinger, T. (2009). *The Ego Tunnel: The Science of the Mind and the Myth of the Self.* New York: Basic Books.

Miłkowski, M. (2016). A Mechanistic Account of Computational Explanation in Cognitive Science and Computational Neuroscience. In *Computing and Philosophy* (pp. 191–205). Cham: Springer International Publishing. https://doi.org/10.1007/978-3-319-23291-1_13

Mitchell, S. D. (2012). *Unsimple Truths: Science, Complexity, and Policy.* Chicago: University of Chicago Press.

Mitchell, S. D., & Dietrich, M. R. (2006). Integration Without Unification: An Argument for Pluralism in the Biological Sciences. *The American Naturalist,* *168*(S6), S73–S79. https://doi.org/10.1086/509050

Niedenthal, P. M., Barsalou, L. W., Winkielman, P., Krauth-Gruber, S., & Ric, F. (2005). Embodiment in Attitudes, Social Perception, and Emotion. *Personality and Social Psychology Review,* *9*(3), 184–211. https://doi.org/10.1207/s15327957pspr0903_1

Northoff, G., & Bermpohl, F. (2004). Cortical Midline Structures and the Self. *Trends in Cognitive Sciences,* *8*(3), 102–107. https://doi.org/10.1016/j.tics.2004.01.004

Northoff, G., Heinzel, A., Northoff, G., Popper, K., Eccles, J., Varela, F., … Churchland, P. (2006). First-Person Neuroscience: A New Methodological Approach for Linking Mental and Neuronal States. *Philosophy, Ethics, and Humanities in Medicine,* *1*(1), 3. https://doi.org/10.1186/1747-5341-1-3

Pezzulo, G., Barsalou, L. W., Cangelosi, A., Fischer, M. H., McRae, K., & Spivey, M. J. (2011). The Mechanics of Embodiment: A Dialog on

Embodiment and Computational Modeling. *Frontiers in Psychology, 2*(5). https://doi.org/10.3389/fpsyg.2011.00005

Qin, P., Di, H., Liu, Y., Yu, S., Gong, Q., Duncan, N., ... Northoff, G. (2010). Anterior Cingulate Activity and the Self in Disorders of Consciousness. *Human Brain Mapping, 31*(12), 1993–2002. https://doi.org/10.1002/hbm.20989

Qin, P., Duncan, N., & Northoff, G. (2013). Why and How Is the Self-Related to the Brain Midline Regions? *Frontiers in Human Neuroscience, 7*(909). https://doi.org/10.3389/fnhum.2013.00909

Quine, W. V. O. (1960). *Word and Object* (2015th ed.). Cambridge, MA: MIT Press. Retrieved from https://mitpress.mit.edu/books/word-and-object-new-edition

Strawson, P. F. (1959). *Individuals: An Essay in Descriptive Metaphysics.* London: Methuen.

Summerfield, J. J., Hassabis, D., & Maguire, E. A. (2009). Cortical Midline Involvement in Autobiographical Memory. *NeuroImage, 44*(3), 1188–1200. https://doi.org/10.1016/j.neuroimage.2008.09.033

4

The Structural Realist Theory of the Self

In the previous chapter, I explained that various scientific (or scientifically informed) theories of the self portray various, and at times conflicting, pictures of the self. None of the pictures seem to provide a wholesome representation of the self. Scientific accounts of the self diverge from the orthodox substantivalist account of the self, but their trajectories are also diverging from one another, and they end up in disagreement without providing a philosophically plausible alternative to the orthodox substantivalist view. In the same vein, in the previous chapter I went out of my way to show how it is that some of the most viable alternatives to the substantivalist theory of the self—for example, Gallagher's pattern theory or Metzinger's eliminativist theory—fail to deliver on their promises because their strategies either are somewhat underdeveloped or collapse into versions of substantivalism. As interesting as various scientific divergences from substantivalism are, we still need to construct a unified

Parts of this chapter are reprinted with the kind permission from Springer Nature. The extracts are taken from two following papers of mine, 2016. "Structural Realist Account of the Self." *Synthese* 193 (12). Springer Netherlands: 3727–3740; 2018. "An Outline of a Unified Theory of the Relational Self: Grounding the Self in the Manifold of Interpersonal Relations." *Phenomenology and the Cognitive Sciences* 1–19.

© The Author(s) 2019
M. D. Beni, *Structuring the Self*, New Directions in Philosophy and Cognitive Science,
https://doi.org/10.1007/978-3-030-31102-5_4

metaphysical alternative to the substantivalist theory. The alternative needs to bring order to the ontological implications of various scientific accounts about diverse aspects of the self.

In Chap. 2, I pointed out that when facing the challenge of underdetermination of metaphysics by many particle physics in the field of philosophy of physics, SR reconceptualises the metaphysics of objecthood. Ontic structural realists such as James Ladyman and Steven French (French, 2014; French & Ladyman, 2003; Ladyman, 1998) have argued that a scientifically informed metaphysics of science should steer clear from object-oriented commitments of the substantivalist metaphysics. Instead, drawing on the preceding work of John Worrall they have introduced a new structuralist approach to metaphysics. According to this new approach, we may let the common mathematical structure of theories guide our attempts at forging ontological commitments. Thus, ontological commitments can be based on the commonality or the common underlying structure of theories. In this fashion, SR makes an improvement on classical scientific realism, which presumes that individual objects (as referents of theoretical terms) are subjects of ontological commitments. SR includes realist commitments to the physical structures in the real world. Like any other philosophical theory, Ontic structural realism (OSR) has won uncompromising critics (such as Psillos, 2001) as well as passionate advocates (such as French and Ladyman op. cit.), but the fact remains that OSR has thrived remarkably in the last few decades. In this chapter, I argue that in the field of the philosophy of selfhood too, we can face the challenge of underdetermination and find a viable metaphysical alternative to the substantivalist theory by advocating a structural realist theory of the self (SRS). Through SRS therefore, I argue that it is possible to reconcile conflicting philosophical implications of diverse theories. In what follows, I flesh out this claim and explain how it is that SRS provides a viable alternative to the substantivalist theory.

I will introduce SRS in Sect. 4.2, before explaining how SRS can address the problem of metaphysical underdetermination in a way that remains beyond the scope of the substantivalist theory of the self. Subsequently, I will specify the difference between *ontologically constitutive* structures of the self and the information-theoretic structures that *represent* the basic structure of the self.

4.1   Informing Philosophy of Selfhood by Sciences

Scientific realism generally aims to develop its epistemological and ontological claims with an eye to what scientific theories say about the world. That is to say, scientific realism advocates realism on the basis of scientific theories' representations of the causal structure of the world. However, not all versions of scientific realism take scientific theories seriously; traditional forms of realism endeavour to domesticate the implications of scientific theories to habituated and a priori metaphysical intuitions (Ladyman & Ross, 2007). OSR dissents from this orthodox approach and advocates a progressive view on the theoretical implications of scientific theories (Ladyman & Ross, 2007). SRS too is a naturalist form of scientific realism which aims to inform its epistemological and ontological commitments on the basis of the results of scientific theories. SRS' commitment to naturalism provides an advantage over the classical substantivalism, which is not in harmony with the new scientific findings about the self. Accordingly, to substantiate the plausibility of SRS—that is, to show that it does really provide a viable alternative to substantivalism—I endeavour to show the underlying structure of the self can be specified not only speculatively but on the basis of some of our best scientific theories of the field. To this effect, I call theories of cortical midline structure (CMS) and predictive processing into play. I build on recent breakthroughs in cognitive neuroscience, firstly to specify the underpinning, embodied neurological structures that realise the pattern of selfhood, and secondly, to characterise these structures at the metatheoretical level, in terms of information-theoretic structures that represent the basic structure of the self. In short, there are various reasons for being optimistic about the prospects of SRS. It possesses remarkable unifying power, and it is in harmony with scientific findings and closely connected with them. On such grounds, it can be argued that SRS provides a viable metaphysical alternative to the substantivalist conception of the self.

4.2   Constructing the Structural Realist Theory of the Self

In Chap. 2, I argued that the standard version of scientific realism—aka object-oriented realism—cannot find a solution to the problem of under-determination of metaphysics by many particle physicists. As I have argued there, modern physics violates some fundamental principles of substantivalist metaphysics such as the Leibnitz's law. Some formulations of quantum physics, such as Bose-Einstein statistics, violate the main tenet of substantivalist metaphysics, which gives ontological priority to individual objects. This is because Bose-Einstein statistics indicates that discrete energy states include distributions of indistinguishable particles, such as bosons. Since bosons are indistinguishable from one another and share all of their state-dependent properties, Bose-Einstein statistics violates the principle of identity of indiscernibles and is at odds with the substantivalist metaphysics. There are other formulations of quantum statistics, for instance, Fermi-Dirac statistics, which are in harmony with the principle of identity of indiscernibles. Thus, there are two ontological packages, one of which retains individual objects, whereas the other formulation denounces them. As I have argued, the orthodox substantivalist metaphysics cannot deal with the disarray that overcomes the ontological consequences of sub-particle physics. Under the circumstances, ontic structural realists such as French (French, 2014; French & Krause, 2010) and Ladyman (1998, 2007) have argued that it is best to find an alternative ontological basis for dealing with the philosophical implications of modern physics. Ontic structural realism (OSR) is the proposed alternative to the substantivalist metaphysics. Unlike substantivalist ontology, it gives ontological priority to structures and relations. Instead of assuming that individual objects are ontologically basic, it reconceptualises sub-particle objects or spacetime points structurally. Thus, OSR as a viable alternative to substantivalism does not seek to make ontological commitments to individual objects but to commonalities or common underpinning structures which lie beneath disarraying formulations of quantum statistics. The commonality could be specified on the basis of the mathematical form of theories. Nevertheless, it is important to notice that, at least according to Steven French's version of OSR, mathematical structures mainly play a representational part. They are invoked to represent physical structures in the real world. Ontic structural realists aim to make ontological commitments to physical structures, which according to some versions of OSR (Esfeld, 2009) can be identified as nomological structures or causal structures in the world. As I say, OSR is a viable alternative to substantivalism. The viability of OSR can be substantiated on account of its remarkable consistency with recent scientific breakthroughs. The epistemology and ontology of OSR are tailored to what scientific theories reveal about the fundamental structure of the world. In this sense, OSR is a fully naturalistic philosophical theory (Ladyman & Ross, 2007). It is also worth mentioning that various advocates of the orthodox or object-oriented scientific realism, such as Hilary Putnam (1975) or Richard Boyd (1980) too, have claimed that scientific realism is a naturalist theory of philosophy. However, to the extent that object-oriented scientific realism could absorb the impact of the scientific developments at all, it has to domesticate them to the habituated metaphysical intuitions (Ladyman & Ross, 2007). In this respect, OSR has the advantage of object-oriented scientific realism, because OSR remains more thoroughly loyal to the spirit of naturalism. OSR lets scientific theories inform epistemology and metaphysics, instead of adopting a priori rigid metaphysical views and then tinkering with the philosophical implications of scientific theories to make them fit in with the object-oriented metaphysical framework.

In Chap. 3, I explained that the theoretical diversity of theories of cognitive science and neuroscience reproduces a form of the problem of underdetermination of ontology of selfhood by neuroscientific theories. Gallagher's pattern theory, for example, presumes that there are multiple self-patterns that realise what we call the self simultaneously. The multiplicity of self-patterns, none of which is necessary or sufficient for the generation of the concept of the self, is enough for reproducing a form of the underdetermination problem. Then there is the divergence between Gallagher's pattern theory, Metzinger's eliminativism, and other accounts such as the CMS theory and the embodied account of the relational self. These scientific accounts of the self have little affinity with the orthodox substantivalist theory of the self. This indicates that we cannot rely on the substantivalist theory to reconcile diversity and overcome the problem of underdetermination. Actually, the opposition between the substantivalist view and recent theories enhances the vicious force of the underdetermination problem. In this chapter, I present the structural realist theory of the self (SRS) as a viable alternative to the substantivalist theory of the self. Unlike substantivalism, SRS is not concerned with the existence of individual objects or primary substances. It provides an alternative metaphysical paradigm, which can take in the philosophical implications of recent theories of cognitive psychology and computational neuroscience and use them to reconceptualise the self as a structural object. In comparison with the orthodox substantivalist theory, SRS is more thoroughly loyal to the results of the scientific theories of the field. Moreover, unlike the substantivalist account, SRS can provide a strategy for addressing the problem of underdetermination of ontology of self by the diversity of recent scientific accounts. According to this proposal, ontological commitments have to be made to commonalities or underpinning structures that lie beneath diverse theoretical descriptions of the self.

SRS is inspired by OSR in the philosophy of science. More specifically, I draw on Informational Structural Realism (presented in Sect. 2.4 of Chap. 2) to introduce SRS. Because recent computational neuroscience and cognitive psychology use *informational*/computational models to explicate the brain's information-processing mechanisms (in the sense that is discussed in Sects. 1.8 and 1.9 of Chap. 1), using information-theoretic structures in the scaffold of SRS will be in order. I will return to this claim later in the final sections of this chapter, and I will elaborate on some scientific theories that describe mechanisms of processing of self-related information in the brain and nervous system. But, right now, the important point is that, given the central role of informational/computational processes in the explication of the functioning of the cognitive system, it is best to develop the structural realist theory of the self in terms of Informational SR (ISR). It is worth mentioning that I do not give way to classical computationalism when I speak of the central role of computational/information processes. As I will explain in Sects. 4.6 and 4.7, I advocate a moderate embodied approach to computational/informational models. I assume that informational structures are embodied in regions of the cognitive system, as well as the whole organism and even its environment. This is in line with my earlier remark in Chap. 2, where I submitted that I am committed to a specific form of informational SR that grounds the informational structures in a cognitive system that can be coupled with the environment (also see Beni, 2019b). I will discuss this point later in the final sections of this chapter. Right now I state SRS in the following terms[1];

> SRS: All else being equal (when we are not facing a case of dissociative identity disorder, etc.), for all scientific and metaphysical purposes, and with regard to all of the instrumental, predictive, and explanatory tasks, our accounts of the self are basically informative about the structure of the informational processing in the system under investigation (e.g. CMS). The informationally regimented structure of the selfhood is epistemically and ontologically fundamental. (Beni, 2016, p. 3732)

Thus, SRS identifies the self in terms of the structure of the self-related information processing in the brain and nervous system. Reconceptualising the self in this way does not indicate that specific self-related features and aspects must be removed from the philosophical account of the self altogether, for it is possible to identify some specific aspects and features non-structurally, in terms of elements that are featuring in the infrastructure of the self. The result would be a thoroughly structuralist account of the self that retains specific aspects of the self and identifies them as structures within structures.

SRS is a non-eliminativist form of SR that retains specific aspects and features of the self, even if we assume that these aspects themselves cannot be reconceptualised structurally. This does not mean that the aspect can be identified non-structurally, because SRS does not assume that specific aspects and features cannot be identified on the basis of their intrinsic properties but in mutual dynamical connection with their underpinning structures. That is to say, non-structural features of the self can be

---

[1] This definition is inspired by Floridi's (2008) statement of ISR. However, although the formulation is inspired by Floridi's work, the notion of informational structure that is used in SRS is different from Floridi's notion. This is because unlike what is the case in Floridi's version of ISR, occasions SRS considers informational structures as embodied informational structures, that is, as structures embodied in the cognitive systems of actual (and usually biological) cognitive agents.

identified in virtue of the specific location that they occupy in the infrastructure of the selfhood. SRS does not eliminate specific aspects and features of the self, but rather embeds them in an informational infrastructure.

I have provided a primary statement of SRS. In the next sections, I shall proceed to show how SRS can address the problem of underdetermination of the self-concept by the diversity of scientific accounts of the self. After doing that and in the final sections of the chapter, I will draw on some state-of-the-art theories of cognitive neuroscience and computational neuroscience to characterise the embodied informational structure that constitutes the self ontologically, as well as the abstract informational structure that represents it.

4.3   A Structural Realist Strategy for Overcoming the Multiplicity of the Self-Patterns

The problem of metaphysical underdetermination can emerge on various occasions. At a very basic level, Gallagher's pattern theory (Sect. 3.3 of the previous chapter) itself gives rise to a vicious form of the problem of metaphysical underdetermination. According to Gallagher (2013), what we call an individual self includes multiple co-existing and loosely related self-patterns, none of which provides a sufficient or necessary condition of selfhood. None of these patterns, in other words, provide the sort of information that can determine the essence of the self. The self is underdetermined by the multiplicity of patterns. On the same subject, the problem of metaphysical underdetermination can receive impact from the conflicting philosophical implications of the pattern theory on the one hand and the eliminativist theory on the other. One of these theories indicates that there are many selves, whereas the other indicates that there are no such things as selves. At another level, the divergence between various philosophical accounts of the self (e.g., substantivalism vs. non-substantivalism) can cause the problem of underdetermination. A substantivalist account delineates the self in terms of a primary substance that endures through changes, retains its identity over time, and is the bearer of the properties, whereas new scientifically informed accounts are not committed to the substantivalist understanding. Is the self a substance or are there no such things as substances? To dissolve the problem of underdetermination at various levels, I argue that a structural realist theory of the self identifies the self in terms of the informational structures that subsume diverse aspects and features of the self. I show how a structural realist strategy will dissolve the problem of underdetermination across various levels. Let me elaborate.

I have canvassed and evaluated Gallagher's pattern theory in Chap. 3. In short, the pattern theory assumes that various contextual, cognitive, conceptual, embodied, and other aspects of what we usually recognise as an individual self actually amounts to various self-patterns that are loosely clustered together. According to Gallagher's original statement, "what we call self consists of a complex and sufficient pattern of certain contributors, none of which on their own is necessary or essential to any particular self" (Gallagher, 2013, p. 3). To account for personal identity, Gallagher claims that various aspects of the self, as being represented by various patterns, are compatible and commensurable, and relate various aspects of the self across certain dimensions. But as I have argued, the original statement of the pattern theory actually fails to deliver on its promise of accounting for the relationship between various aspects and elements of the self as being represented by various self-patterns. More recently, Gallagher and Daly (2018) have reemphasised the role of dynamical relations in interrelating dynamical patterns that form an individual person.

According to Gallagher and Daly, the set of dynamical relations could be characterised both at the level of neural dynamics and extra-neural (embodied and enactive) dynamics. Thus, their project aims to specify interrelations between various self-patterns in three significant ways. The relations can be specified in a negative way on the basis of their interruption in pathological situations. They could be specified positively on the basis of the narrative component of a self-pattern, given that "a self-pattern is reflectively reiterated in its narrative component" (Gallagher & Daly, 2018, p. 4). Finally, Gallagher and Daly suggest specifying the relationship between various self-patterns on the basis of the neuronal structures forged through predictive processing in the brain. I do not think

Gallagher and Daly's rejoinder is completely consistent with Gallagher's (2013) original statement of the pattern theory. According to the original statement, the narrative self-pattern or the neuronal self-pattern are just two specific self-patterns amongst numerous others, and they cannot take a privileged status to account for other aspects of the self as being represented by various self-patterns. In addition to previous discussion on these points, I have to add that Gallagher and Daly's recent paper makes a great improvement on Gallagher's original pattern theory. This recent improvement has a pleasant structuralist ring to it. According to them, from the perspective of the pattern theory of the self, "it is not the case that one needs to add something extra to the specific pattern dynamically formed by the interrelations of the various elements in order to be able to identify the coherency or structure or ordering behind the diversity; rather the coherency, structure or ordering will be reflected in the pattern that emerges from these interrelations" (Gallagher & Daly, 2018, p. 4). I find this statement in line with an earlier statement of SRS in a previous paper (Beni, 2016).

The structuralist tendency that is aligned with pattern theory is especially observable in Gallagher and Daly's attempt at specifying the relations between self-patterns in terms of dynamical neural connections. They have drawn on recent theories of computational neuroscience, for example, the Free Energy Principle (FEP) and predictive coding, in order to specify the requisite dynamical neural connections. I will speak expansively about FEP and its role in regimenting the structure of the self (in my view, at a meta-level) in Sect. 4.6. Right now, I point out that despite taking great strides towards a structuralist conception of the self, Gallagher does not develop the pattern theory into a fully structuralist theory of the self. That is to say, his interesting engagement with the structuralist theme is not evolved into a well-posed metaphysical alternative to the orthodox substantivalist metaphysics of the self. He only draws attention to the significance of structural relations. For example, he submits that "the coherency, structure or ordering will be reflected in the pattern that emerges from these interrelations" (Gallagher & Daly, 2018, p. 4). Such statements are not developed into a well-posed alternative to the substantivalist discourse, which presumes that the self is a substance that endures over time and is the bearer of properties. Unless we can offer a viable alternative to the substantivalist view to support Gallagher and Daly's advanced version of the pattern theory, the philosophical plausibility of their view cannot be established. There are just diverse self-patterns with only loosely set connections to each other. The diversity of patterns underdetermines the ontology of the self.

SRS, as being introduced in the previous section, provides an alternative paradigm to substantivalist metaphysics. It presumes that the self is the underpinning structure that relates various self-related aspects and features, which identify with what Gallagher has called self-patterns in Gallagher and Daly's theory. Substantivalism gives metaphysical priority to individual objects that feature in the structure. In contrast to substantivalism, SRS presumes that the self is the structure, and the self-structure has ontological priority over the entities that inhabit the structure. The self is the structure that lies beneath various self-patterns and realises them. And this commonality or the shared structure of various self-patterns of self-aspects is the right subject of epistemological and ontological commitments of a realist theory of the self.

Although it might be possible to construe self-patterns as individual objects, I strongly suspect that pattern theory does not conform to the substantivalist discourse, and it is best to understand self-patterns as sets of dynamical relations—self-patterns are *patterns* after all. This means that Gallagher's theory does not give priority to individual objects and it is compatible with structuralism. If so, it would be best to understand the pattern theory in terms of SRS which provides a viable metaphysical alternative to substantivalism. According to SRS, the self is the infrastructure that subsumes various self-patterns. Self-patterns or various aspects of the self are featuring in the infrastructure of the self. Aspects and features of the self (i.e., self-patterns) could be identified mainly by virtue of their location within the infrastructure of the self. This reading can account for the integration or unity of the self without collapsing into substantivalism. The underpinning structure that subsumes various self-patterns actually unifies them, without indicating that various aspects are modes or attributes of a substance-like entity. This is the same as saying that SRS could dissolve the underdetermination of the self-concept by the multiplicity of self-patterns, by submitting that the subject of epistemological and ontological commitments is the shared structure that lies beneath the tide of diversities. It will still be incumbent on me to characterise the underlying structure of the self, but before doing that, in the next section, I shall explain how a structuralist strategy could also reconcile Gallagher's pluralism to Metzinger's eliminativism, and thereby dissolve the problem of underdetermination at another level.

4.4   Overcoming the Eliminativist-Pluralist Dichotomy

While pattern theory is leaning towards pluralism—it identifies the self in terms of multiple and simultaneously existing but loosely integrating self-patterns—Metzinger's theory eliminates the self from the ontology. These two approaches are at odds, and their rivalry leads to a state of underdetermination of our ontological conception of the self by scientific or scientifically informed theories of self-related aspects and properties. Either an individual person is a cluster of multiple self-patterns none of which is singularly sufficient or necessary for underpinning the self, or there are no such things as selves at all. Neither scientific evidence nor philosophical argumentation can corroborate one of these two approaches to the self and reject the other. We do not need to go out of our way to explain that the substantivalist view cannot deal with the situation. Both Gallagher's and Metzinger's theories are diverging from the substantivalist view. Neither of them assumes that the self is a primary Cartesian substance, nor that the self is an indivisible unit that bears the properties and endures through the changes. Thus, both theories diverge from substantivalism. However, as I have argued, neither of them provides a metaphysically viable alternative to substantivalism. SRS, on the other hand, is a viable alternative to substantivalism which could address the problem of underdetermination that raises through the rivalry between pluralism and eliminativism. Below, I shall unpack this remark.

SRS offers to account for the epistemological and ontological aspects of the self in terms of informational structures. From the epistemological point of view, what we can *know* about the self is mainly based on its informational structure. From the ontological point of view, the self *is* the shared infrastructure that underpins the class of self-related informational structure. According to SRS an individual person does not identify with an orthodox substance, but can only be identified relationally, and within the context of a specific class of self-related informational structures. To dissolve the problem of underdetermination of the self-concept by diverse theoretical accounts of the self, I suggest that the fundamental informational structure of the self is the commonality that underpins the pluralist and eliminativist accounts of the self. As soon as we dispense with the substantivalist theory and submit that the self could be identified in terms of the underpinning common structure that subsumes pluralist and eliminativist accounts, the problem of underdetermination dissolves.

I have already explained how a structural realist strategy could support pattern theory, namely, by constraining the multiplicity of self-patterns by the set of dynamical relations that realise infrastructure of the self (see the previous section). SRS gives ontological priority to the infrastructure, and it assumes that specific self-patterns could be identified by virtue of their location in the underpinning structure of the self. We can identify a specific self-pattern relationally and on account of its relationship with other self-patterns that constitute a specific self-pattern. The same underpinning structure that integrates various self-patterns can also subsume Metzinger's self-model and reconcile it to Gallagher's pattern theory.

According to Metzinger, the self-model realises the basic structure of minimal and embodied aspects of the self. In the same vein, the self-model underpins the experience of body ownership—that is, the feeling that one's self is embodied with a specific body—as well as the property of perspectivalness—that is, the thesis that a self occupies the first-person point of view (Lenggenhager, Tadi, Metzinger, & Blanke, 2007; Metzinger, 2003). These aspects of the minimal and embodied self can be identified as specific self-patterns, for example, minimal self-pattern, embodied self-pattern, and situated self-pattern. Accordingly, the underpinning infrastructure of the self can subsume all of these aspects and unify them. That the informational structure of the self lies beneath both Gallagher's self-patterns and Metzinger's self-models in this vein unifies them. The common underlying structure includes sets of dynamical relations that are supposed to connect various self-patterns in the context of Gallagher's theory. Additionally, the underlying structure identifies with the form of Metzinger's self-model which is supposed to include the main aspects and elements of the traditional self. However, unlike both of these theories, SRS emphasises the point that the basic structure of the self itself is the subject of ontological commitments. It may be remarked that "mineness", the sense of ownership, and the sense of agency, which seem to be central to phenomenal selfhood, do not identify as structural notions. This assertion does not contradict the tenet of SRS, according to which the basic structure of the self underpins various aspects and elements of selfhood. This is because SRS is a non-eliminativist form of information SR, and it does not eliminate non-structural elements such as the sense ownership, the sense of agency, and so on. These notions could feature in the basic structure of the self, to the extent that we could account for their *connection* with the basic structure of the self (for neuroscientific accounts of this connection see Seth & Tsakiris, 2018; Williford, Bennequin, Friston, & Rudrauf, 2018).

Be that as it may, the self-structure underpins Metzinger and Gallagher's respective theories. Metzinger's theory submits that there are no such things as selves in the world. However, by speaking of the self-model, Metzinger retains a substitute or replacement that could effectively act as the self in all scientific and philosophical contexts. He merely eliminates the substantial self but retains a structured self-model. But Metzinger's self-model does not provide a viable alternative to the substantivalist theory of the self, because it cannot account for the relationship between various aspects of the self without succumbing to substantivalism. SRS offers a well-established metaphysical framework whose philosophical claims have already been accredited in the field of philosophy of science. SRS dispenses with the substantial self without endangering the integrity and unity of the self. The underpinning structure of the self preserves the unity of the self by subsuming various self-patterns of Gallagher's theory as well as various aspects of Metzinger's self-model. It vouchsafes the existence of meaningful relations between various aspects of the self because the self is identified in terms of the underpinning dynamical relations that relate various aspects of the self. In the same vein, SRS can address the problem of metaphysical underdetermination because it indicates that the self is the basic structure that underpins diversified accounts of the self and its aspects. Epistemological and ontological commitments are made to the common underpinning information-theoretic structure that subsumes diverse theoretical concepts manifested by self-pattern or self-model.

The substantivalist theory of the self cannot make sense of the confusing theoretical diversities, still much less dissolve the state of underdetermination that is caused by them without domesticating scientific finding to a priori insights. SRS, on the other hand, provides a viable metaphysical alternative which is firstly in harmony with the results of scientific accounts of the self, and secondly subsumes theoretical diversities and reconciles them. In addition to its harmony with the philosophical implications of scientific theories of the field, SRS's great unifying power provides further ground for being optimistic about its philosophical viability.

Thus far, I have argued for the viability of SRS by elaborating on its capacity for addressing the challenge of metaphysical underdetermination and by remarking on its unifying power. In the remainder of this chapter, I shall provide further details that cement the plausibility of this theory. I shall engage in characterising the *nature* of the underpinning structures.

4.5 The Role of Information-Processing Mechanisms

Thus far, I have argued that SRS can dissolve various forms of underdetermination. It provides a viable alternative to substantivalism. To consolidate this claim, I must elaborate on the nature of the underpinning structure of the self, that is, the shared self-structure that unifies theoretical diversities. In the previous section, I have begun to suggest that the self can be identified in terms of structures realised through the processing of self-related information in the brain and nervous system. In this section, I will specify the embodied informational structure of the self with the requisite details—this structure is embodied in the sense that it is implemented in the brain and nervous system of organisms that have selves. In this vein, I embark on providing necessary details about the underlying mechanisms that contribute to the processing of self-related information in the cognitive system. The point is not only elucidatory, but it is also important from a metaphysical point of view. I explicate the metaphysical component of SRS, at least partly, by suggesting that SRS is ontologically committed to the embodied structure of self-related information processing in the actual mechanisms in certain regions of the brain and nervous system. To be more precise, I will elaborate on the role of cortical midline structures and the Mirror Neuron System, as the neuronal mechanisms that implement the basic structure of the self. Here, the general insight is that the embodied structure of the self can be specified on the basis of these underpinning mechanisms that embody the self-structure.

4.5.1 Cortical Midline Structures

As I explained in the previous chapter, the cortical midline structure theory of the self provides an alternative to both Gallagher's pattern theory and Metzinger's self-model. The CMS theory explicates the self in terms of the processing of self-related information in cortical midline structures. Presently, I am not concerned with the plausibility of the CMS theory as a viable alternative to Gallagher's and Metzinger's theories. I am referring to cortical midline structures as a system that could embody the information structure of the self in collaboration with some other systems, for example, the Mirror Neuron System. Thus, I do not defend the CMS theory as an alternative to pattern theory or eliminativism. The same structuralist strategy that has been used to overcome the problem of underdetermination in the context of theories of Gallagher and Metzinger could be extended to address the problem of disagreement between the CMS and other theories. Instead of proceeding to repeat this strategy to reconcile the CMS theory to other theories, in this section I will explicate the embodied mechanisms of processing of self-related information in terms of the operation of CMS. This demonstration—that is, how CMS implements the structure of the self—helps us to clarify a part of the ontological implications of SRS, by saying that SRS makes realist ontological commitments to informational structures that are embodied in CMS (and some other regions). In short, the brain's cortical midline structures partly embody the basic structure of the self. I do not assume that CMS embodies the self-substance; such an ambitious claim could hardly be substantiated. My claim rather is that CMS's mechanisms implement the informational structure of the self in the sense that is at issue in SRS.

CMS is a part of the Default Mode Network. It consists of several brain regions, including the medial prefrontal cortex, anterior cingulate cortex, and posteromedial cortices (Northoff & Bermpohl, 2004; Qin, Duncan, & Northoff, 2013). The processing of self-related information in CMS realises the fundamental structure that underpins various aspect and features of the self. These aspects and features identify with self-patterns in Gallagher's theory or as aspects of the self-model according to Metzinger's theory. Processing of self-related information in CMS provides a neuronal base for integration of minimal, embodied, and narrative aspects of the self, as well as its experiential aspects and consciousness (I will speak more about phenomenal aspects in the next chapter). CMS partly integrates the autobiographical and emotional aspects of the self and connects them with the first-person perspective. I say "partly" to indicate that the integrative power of CMS is limited. Later, I show that to account for unification between various aspects of the self (and its relationship with its environment) we are in need of a more powerful unifying framework (that will be presented in terms of the Free Energy Principle). However, CMS underpins an experimentally viable account of a class of cognitive mechanisms that embody various aspects of the self. And there are good experimental reasons for assuming that CMS is concerned with the processing of self-related aspects of the self. Self-related cognitive tasks, such as thinking about oneself or one's close friends, are accomplished at the neuronal level through the activation of patterns of processing in CMS (Northoff et al., 2006).

CMS is encompassed by the Default Mode Network that represents the activity of the resting state of the brain, which is usually measured when the subject is not engaged in any goal-directed task. The brain's resting state activity and the Default Mode Network represent the brain's default cognitive template. The brain's regions associated with the Default Mode Network have the highest baseline metabolic activities in the resting state. However, when the subjects are required to engage in cognitive tasks, the activity of the Default Mode Networks will be attenuated (Qin et al., 2013). Being in the resting state mode does not mean that the brain is resting or is engaged in simple overhaul activities such as neuronal repair. The brain's resting state activity is mainly concerned with the processing of self-referential thought and self-related information (Qin & Northoff, 2011; Weiler, Northoff, Damasceno, & Balthazar, 2016). Resting state activity realises the patterns of thinking about one's traits, retrieval of the episodic autobiographical memory, and processing of other dispersed pieces of information about oneself.

There are also negative reasons for presuming that the resting state activity realises patterns of self-relating thinking. Lesions in the Default Mode Network are correlated with disturbance of the temporal continuity of the consciousness or the stability of the self from the first-person perspective (Northoff, 2014a). This provides further (negative) evidence for the existence of a correlation between CMS and reflective aspects of the self. Although CMS realises the structure of reflective aspects of the self, it is not exclusively concerned with realising reflective aspects. As I will explain in Chap. 6, CMS also engages in the processing of social and ecological information (Apps & Tsakiris, 2014, p. 93). I shall speak about the neuronal mechanisms that underpin the social aspects of the self shortly, but right now, suffice it to say that CMS also engages in processing information concerning the relationship between oneself and the social and ecological context. This means that while the mechanisms of CMS are self-related, they are not exclusively self-specific. CMS embodies the informational structure of the self, in the sense that it includes neuronal mechanisms that jointly engage the processing of self-related information. At the same time, although CMS is mainly engaged with the processing of self-related information—that is, it mainly realises the reflective aspects of the self—it can also be related to the social and ecological aspects of the self. On the one hand, CMS mechanisms embody the informational structure of the reflective aspects of the self. On the other, the spatiotemporal structure of the brain's activity represents the causal structure of the environment and society. Putting these two brain functions together, it appears that the brain structure could embody the informational link between the subjective experience of the self and the representations of the social and ecological environment (Beni, 2019a;

Northoff, 2014b). It is possible to account for the social aspects of the self at least partly on the basis of mechanisms of CMS, by saying that the brain's capacity to represent the structure of the (social and ecological) environment is at least partly based on the information-processing mechanisms of CMS. This does not mean, though, that CMS is the only neuronal system that realises the social and ecological aspects of the self.

Aside from CMS, and perhaps more conspicuously than it, the Mirror Neuron System (MNS) includes the set of neuronal mechanisms that provide a neurophysiological base for the social aspects of the self. I will elaborate on the social aspects of the self in Chap. 6. But in the next section, I will explain that the mechanisms of processing of self-related information in CMS are coupled with mechanisms of processing of social information, that is, mechanisms of the Mirror Neuron System (MNS). The self includes both reflective and social aspects. Thus, our account of the neuronal mechanisms that underpin the processing of self-related information must include both CMS and MNS.

4.5.2   The Mirror Neuron System

The classical representationalist models of cognition and perception assume that mental representations are processed as amodal abstract data structures. Embodied theories challenge this assumption and emphasise the role of embodied mechanisms and sensorimotor system in cognition. When applied to the theories of social cognition, the embodiment may indicate that our understanding of other people's mental states is unreflective and direct and it does not mandate classical representational models. When we observe an action, our motor system automatically and unconsciously simulates the performed action, albeit without actually replicating the action in question (Gallese, 2014). Grounded mechanisms of cognition aim to describe the mechanisms of embodied simulations. Embodied simulations underlie the possibility of understating how someone else executes an action without actually replicating it. Neurologically, embodied simulation—which underlies social cognition—is a result of the coupled processing mechanisms of motor and control systems. The theory of embodied simulation draws a connection between the experience of bodily modes of oneself on the one hand and awareness of others and their intentional and goal-directed actions on the other hand (Gallese, 2005). The discovery of mirror neurons and their function provided a new foundation for developing the theory of embodied simulation, by explaining how the brain's MNS contributes to decoding the others' sensations, emotions, and intentions directly. According to this theory, mirror neurons are multimodal motor neurons that populate ventral premotor area F5, area FF, and the inferior parietal lobe of the brains of macaque monkeys. These neurons are active both when the monkeys engage in object-related action or/and when they observe others' (monkeys' or humans') actions (Gallese, Keysers, & Rizzolatti, 2004; Umiltà et al., 2001). The MNS is sensitive to partly concealed actions and motor goals that cannot be fully observed, as well as fully observable actions. Although the research on mirror neurons used monkeys as subjects initially, MNS operates almost in the same way across humans' brains and macaques' brains. In humans, the observation of the goal-directed actions of others is associated with the activation of the premotor and parietal areas of the human brain (Gallese, Eagle, & Migone, 2007). The Mirror Neuron Systems of humans, too, contribute to the imitation of simple acts, learning complex unpractised motor acts, as well as understanding others' actions and intentions. Also, it is possible to trace the activity of the Mirror Neuron System both in new-born children and in adults (Wohlschläger, Haggard, Gesierich, & Prinz, 2003). The model of social cognition that is based on the operation of MNS explains mindreading,[2] affective mechanisms, action understanding, and empathy. The MNS theory provides a viable explanation of how the observation of actions of others invokes the unconscious and automatically simulated re-enactment of the same action in the observer. It provides a basis for a meaningful account of the goals and purposes that motivate the actions of others (Gallese, 2003).

---

[2] The embodied approaches to cognition provide an alternative to the classical forms of cognitivism and representationalism (for explanation of the embodied approach see Sects. 4.6 and 4.7 below). Accordingly, the embodied approach denies that social cognition mandates mindreading or the capacity to represent others' mental states and reacting to them (Hutto, 2015; Spaulding, 2012). Gallese's theory includes elements from the embodied approach too, but he accounts for understanding other people in terms of "embodied simulation" of their mental states. Accordingly, and in line with simulation theory, he uses the Mirror Neuron System to explain mindreading and empathy (Gallese, 2003, sections 5 and 6). I will speak about embodied simulations more in Chap. 6. For the time being, it may suffice to say that I find this approach in line with the version of moderate embodied approach that I am advocating in this book (see Sects. 4.6 and 4.8).

Vittorio Gallese (2003, 2005) has developed an interesting theory of interpersonal relations and intersubjectivity on the basis of the operation of MNS. To be more precise, Gallese's theory of interpersonal relations draws on MNS mechanisms to integrate the account of bodily awareness with the account of basic forms of social understanding. On such grounds, it could be claimed that mechanisms of MNS underpin social aspects of the self. I will speak more about Gallese's theory in Chap. 6, and I will explain that Gallese's work (and some other similar studies) provides a nice base for a structuralist theory of social aspects of the self. This section, however, is concerned with identifying the embodied informational structure of the self, rather than the social aspects of the self. Accordingly, in the remainder of this section, I elaborate on the relationship between CMS and the MNS, so as to substantiate the point that these two systems jointly form the mechanism that underpins the processing of the self-related information in the brain.

I show that CMS's neural activity weaves into the activity of the mirror neuron network. I rely on Uddin, Iacoboni, Lange, and Keenan's (2007) review of the neurological literature to substantiate the point. There seems to be an essential difference between the respective functions of CMS and MNS. CMS is mainly (though not exclusively) concerned with the processing of self-related information and realises the reflective aspects of the self. MNS, on the other hand, is engaged with the processing of patterns of information about the others-to-self relationship. These patterns are necessary for understanding the physical actions of others. MNS supports motor simulation and underpins social cognition, whereas CMS is engaged with the processing of self-evaluative and self-related simulations. Within this context, Uddin et al. endeavour to reconcile two separate fields of research that deal with CMS and MNS. Their proposal could underpin a unified account of the reflective and social aspects of the self (Beni, 2019a), for the coupled system constituted by CMS and MNS embody the informational structure of the self.

According to Uddin et al., because both CMS and the mirror neuron networks engage in processing of self-related and other-related information, it can be assumed that they form a dynamical relationship with one another. Uddin et al. rely on various experimental findings to support this claim, drawing on previous experimental studies (Giambra,

1995; Mitchell, Macrae, & Banaji, 2006; Rizzolatti & Luppino, 2001) to argue that there is a significant relationship between the precuneus (a major portion of the CMS) and the inferior parietal lobule (the posterior component of MNS). There is also a connection between mesial frontal areas and the inferior frontal gyrus, which indicates that CMS can engage in motor simulation. Uddin et al. refer to fMRI-based studies of self-recognition patterns, according to which there is a dynamical relationship between self-recognition and mirror neuron areas in the right hemisphere of the brain. This adds up to the assumption of a significant relationship between the neural underpinning of reflective aspects of the self (in CMS) and the neuronal underpinning of imitation and action observation (in the MNS) (Uddin et al., 2007, p. 156). I will speak about the MNS and its role with the social aspects of the self in Chap. 6.

Because there are dynamical patterns of neuronal activity that inter-connected mechanisms of CMS and the MNS, it could be claimed that the common embodied informational structure that subsumes reflective and social aspects of the self is the subject of ontological commitments of SRS. This was the main motive for arguing that CMS and the MNS jointly form the mechanism that underpins the embodied informational structure of the self. In this fashion, I specified the embodied informational structure that is the subject of SRS's epistemological and ontological commitments at the neuronal level.

It is also possible to formulate the structure of the self at the level of abstract formal mathematical relations. In the next section, I draw attention to a family of blooming theories of computational neuroscience that can be subsumed under the general category of the Free Energy Principle, and I show how FEP can be used to present a mathematical characterisation of the informational structure of the self too.

4.6   The Free Energy Principle and Meta-Theoretical Informational Structures

The informational structure of the self can be characterised, as an embodied structure, at the level of neurophysiological mechanisms. However, it is also possible to regiment the informational structure of the self along the lines of predictive processing and the Free Energy Principle (FEP). FEP-based characterisation of the informational structure of the self could be too abstract to identify the neurophysiological mechanisms that underpin the self. However, it can provide a comprehensive theoretical framework for representing the self, its various aspects, and its representational powers with remarkable mathematical precision.

As I have remarked previously in this book, Gallagher, too, suggests that it is possible to model the relationship between diverse self-patterns in terms of dynamical neural connections that are described by predictive coding and FEP (Gallagher & Daly, 2018). While I appreciate the insight behind Gallagher's suggestion, I do not think his reference to FEP can provide a viable account of the unity of the self in the context of the pattern theory. For one thing, as I have explained before, Gallagher's proposal does not evolve into a viable alternative to substantivalism. SRS, on the other hand, is an alternative to substantivalism that can build upon the neuro-computational theories such as predictive coding and FEP to characterise the informational structure of the self. Another reason for my disagreement with Gallagher is more specific; on several occasions, Gallagher has suggested that he understands predictive coding along the lines of theories of embodied cognition (Gallagher, 2014; Gallagher & Allen, 2016).

I appreciate Gallagher's approach. However, while it is possible to advocate an embodied understanding of predictive coding and FEP (see Beni, 2018a), as I will explain shortly, there are reasons for assuming that the patterns that can be regimented in terms of predictive coding and FEP include both representational (in the computational sense) and embodied aspects, instead of purely embodied aspects. This means that FEP-based patterns could be used to draw connection between abstract models of aspects of the self and the embodied underpinning mechanisms of those aspects. In saying this, I follow Pezzulo et al.'s (2011, 2012) view on the possibility of an alliance between computational and embodied approaches. I will return to this topic in the next section. To be clear, here, I submit that I use FEP and predictive coding not only for specifying the set of dynamical relations between various self-patterns or aspects of the self but also for the more fundamental task of identifying the basic informational structure of the self at an abstract representational level (I have already specified the embodied informational structures in terms of coupled mechanisms of CMS and MNS).

According to a promising theory of contemporary computational neuroscience, the brain's main activity consists of predictive processing. Predictive processing theory (PP) aims to explain cognition, memory, action, learning, and perception by invoking a very few computational principles (Clark, 2016b; Hohwy, 2013; Huang & Rao, 2011). PP holds that the process (of cognition, perception, learning, etc.) begins by generating internal models randomly. After generating internal models, the brain tries to apply its models to the world through top-down processing and error-reduction mechanisms. The top-down stream of the brain's predictions meets the upward coming stream of prediction errors. Perception and cognition are the results of the brain's attempt at reducing the prediction error. This means that the brain's attempt at capturing and representing the causal structure of the world consists in minimising the discrepancy (i.e., the prediction error) between the brain's expected inputs and the actual inputs that represent the hidden causes of perceptions via neural computations which approximate Bayesian inferences. The brain uses empirical Bayesian mechanisms to infer the causes of its sensory inputs, given the likelihoods (or predicted inputs), where predicted inputs are generated by the predictive or generative models of the causes.

Karl Friston has worked out the biological basis of PP and articulated it in terms of the Free Energy Principle (FEP) (Friston, 2010; Friston & Stephan, 2007). FEP aims to explain in biologically viable terms why organisms tend to minimise their prediction error. According to FEP, the free energy is an information-theoretic measure which bounds or limits the surprise on sampling some data, given a generative model. The surprisal is identified in terms of the negative log-probability of an outcome, where entropy is the average surprise of an outcome. A generative model is a probabilistic model of the dependencies between causes and data, from which samples can be generated (Friston, 2010, pp. 1–3). According to this theory, maximising survival is proportional to minimising the free energy which is the same thing as the prediction error. There are fine evolutionary reasons for assuming that there is a relationship between minimising prediction error and maximising survival. For evolutionary reasons, organisms must maintain their internal states within bounds delineated by FEP and reduce their internal models' prediction errors to save their energy. This enables the organisms to resist the dispersing effect of the environment, endure, and survive within an environment with changing features. Organisms can reduce their prediction error (or the amount of the surprisal) by invoking active inferences, that is, either by changing their internal states in order to reduce their prediction error or alternatively by changing the environment in order to make the actual input conformable to their predictions (Friston & Stephan, 2007). It is also worth mentioning that FEP, predictive coding, and predictive timing, as well as active inferences, rely on the mechanisms of propagation of efference copies from the motor to sensory cortices. Top-down processing itself could be construed as an efference neural operation, where the interface between top-down generative processes and upward coming actual stimuli could be construed in terms of the encounter between efference copies and afferent information (Arnal & Giraud, 2012, p. 393).

Be that as it may, FEP and PP could explain not only perception and cognition in the context of exteroception but also introspection and the processing of self-related stimuli. This point has been extensively discussed in a number of recent studies (Apps & Tsakiris, 2014; Hohwy & Paton, 2010; Limanowski & Blankenburg, 2013; Petkova & Ehrsson, 2008; Seth, 2013). According to these studies, some of the basic aspects of the self, for example, the embodied aspects of the self, could be modelled on the basis of mechanisms of predictive multisensory integration of self-related (visual, tactile, proprioceptive) stimuli as described by FEP and PP (Seth, 2013, pp. 565–566). This means that FEP and PP could generate a model of the (organism's) self. In the case of the embodied aspect, for example, the organism's model of itself is based on its predictive representation of its own body—that is, the location of the body, its morphology, and so on. As Seth has submitted, the predictive model of the self "must extend beyond subjective feelings to integrate interoceptive and exteroceptive signals across multiple levels of self-representation. Particularly significant is the representation of those parts of the world perceived as belonging to one's own body" (Seth, 2013, p. 569). In the same vein, it has been argued that self-recognition is a result of the brain's attempt at minimising its free energy. For the predictions to be correct, the agent must place itself in regions of the environment that are most predictable (Apps & Tsakiris, 2014, p. 86; Fotopoulou, 2012). In all, this approach provides a probabilistic basis for accounting for the mental representation of the physical properties of one's self.

The FEP-based account of the self is in harmony with embodied theories because it constructs its account of the self and its phenomenal aspects on one's conception of one's own body. One's body, according to this approach, has the highest probability of being "me"—whereas the likelihood of other objects invoking the same sensory inputs is relatively lower (Apps & Tsakiris, 2014, p. 88). Given the reliance of modelling of self-related aspects on the position of one's body, it might be assumed that FEP-based theory of the self belongs to the theories of embodied cognition. This does not need to mean, however, that the predictive processing and the free energy system themselves are embodied neurophysiological mechanisms. The fact that FEP and PP's formalism could subsume the embodied aspect of the self does not indicate that FEP and PP themselves can be identified with embodied mechanisms. Rather the point is that FEP provides a unifying framework with great expressive power and that this vast framework could be used to model a number of different things, the embodied aspects of the self included. I will develop this theme in the next section.

Be that as it may, the FEP-based theory of the self could account for self-identification. The same theoretical framework could be invoked to account for how one recognises one's own self on the basis of the congruent corollary discharge, vestibular, somatosensory, and interoceptive information evoked when one sees her own face on a reflective surface (Apps & Tsakiris, 2014, p. 88). This approach is consistent with (and is based on) previous views on the connection between self-awareness and the physical features of the embodied self (Seth, Suzuki, & Critchley, 2012). It is even possible to use FEP-based theory to account for cases of bodily illusions.

As I remarked before (when explaining Campbell's point), motor efferences are copies of motor commands. Predictions of multisensory consequences of actions are generated by these motor commands. Reafferences are the actual sensory consequences of actions (Apps & Tsakiris, 2014, p. 86). This means that self-specific predictive processing consists of an interplay between introspective efference and reafference streams. The information flowing down through the hierarchy is in the form of expected inputs or predictions about the consequences of events (Apps & Tsakiris, 2014, p. 87). Representation is a result of the encounter of top-down predictions and the bottom-up stream of actual inputs. An empirical (or variational) Bayesian framework can model this encounter. In this vein, Apps and Tsakiris built upon Friston's theory to argue that FEP could account for disturbances in self-identification and self-recognition by the convergence of the multisensory nodes at the higher levels of the hierarchy which store abstract information about the self. In the same vein, bodily illusions could be explained on the basis of the top-down mechanisms of generating models and reducing prediction errors. This is because FEP and PP can explain self-identification and self-location as abstract models encoded into top levels of the hierarchical processing in the multisensory association areas (Apps & Tsakiris, 2014, p. 88 ff.). The top-down effect of multisensory processing could minimise the amount of surprisal in a self-recognition process. Within this context, bodily illusions could be explained in terms of the plasticity of the mechanisms of error prediction and FEP. For example, a local bodily illusion such as the Rubber Hand Illusion[3] could be explained in the following way: the discrepancy between the top-level models of bodily reference frames and the content of the upward-going actual visual models would be explained away by assuming that the touched rubber hand is the person's real hand (hence the illusion). The illusion occurs because in order to minimise the prediction error (or keep FEP low) the brain shifts the participant's perception of the location of the participant's real hand to a spatial location closer to the rubber hand than its actual location (Apps & Tsakiris, 2014, p. 89).

The above-mentioned explanation applies to various cases of full-body illusions (Blanke, Landis, Spinelli, & Seeck, 2004). For example, in the case of heautoscopy, in order to retain the generative model according to which there is a connection between the experience of self-location and

---

[3] In the experimental setting, the experimenters put a rubber hand in front of the subject and hide her real hand from her view. While the subject observes the tactile stimulation of the rubber hand, she will be stimulated on her real hand simultaneously. The temporal synchronicity of stimulations on congruent specular locations on the respective real and rubber hands induces the feeling of ownership over the rubber hand in the subject (Apps & Tsakiris, 2014, p. 89).

Minimal Phenomenal Selfhood (MPS), the brain shifts to the experience of being sliced into two selves or being two persons at the same time. The brain's model of the self at the higher level of the hierarchy misrepresents the illusory body or body part (e.g., the rubber hand) as a part of the person's own body (or identifies it with the person's own full body). The counterevidence—for example, one's knowledge that the rubber hand is not mine, or the neurological fact that the actual sensory input does not fit with the illusory input, and so on—would be dismissed on account of being too surprising to be reconciled with the generative model. In the cases of full or partial body illusions, the generative model of the other (illusory) body being my own would get a higher likelihood than the model that ascribes my own real body to me. The actual input, which indicates that the illusory body is not mine, would be suppressed by the generative model and the efference copying mechanisms. And there are neurological studies that indicate that the parts of the brain—for instance, temporoparietal junction (TPJ), the superior temporal sulcus (STS)—that are engaged with the processing of multisensory self-specific stimuli are also involved when the subject is experiencing multisensory bodily illusions (Apps & Tsakiris, 2014, p. 93). All of these indicate that FEP can underpin a viable theory of selfhood. Accordingly, it could be argued that the self-structure can be characterised in terms of Bayesian structures that are presented by FEP.

Let us recap. Apps and Tsakiris (also Limanowski & Blankenburg, 2013) take great pains to account for partial and full-body illusions on the basis of FEP. In this vein, they argue that "the driving effect for each illusion is an increase in the probability that the other object (a face, voice or body part) will be represented as part of the body and a decrease in the probability that one's actual body will be represented as 'self'" (Apps & Tsakiris, 2014, p. 90). They point out that during the illusory experience, there are even physiological changes (e.g., the decrease in the temperature of the real hand, in the Rubber Hand Illusion) which evince that the real limb is rejected and that there is a decreased likelihood that it is a part of one's body. This is because the representation of the self (self-identification, limb location, etc.) takes place under the rubric of FEP and its probabilistic top-down mechanisms. The fact that FEP and PP can be used to account for the case of bodily illusions provides further grounds for being optimistic about the prospects of an FEP-based theory of the self. Thus far, in this section, I have shown that FEP can model the embodied and phenomenal aspects of the self adequately (I will speak more about the role of FEP in explicating phenomenal aspects in Chap. 5).

As I remarked earlier in this section, there are both representationalist (Hohwy, 2013; Hohwy & Paton, 2010) and embodied-enactivist interpretations of FEP (Bruineberg, Kiverstein, & Rietveld, 2016; Clark, 2016a). I do not want to engage the debate over the right philosophical construal of FEP here. I generally assume that both parties could devise arguments to defend their views viably enough (Beni, 2018a). What is important is that, as parallel alternative interpretations indicate, FEP provides a unifying representational framework as well as an account of embodied mechanisms of cognition. We can use the FEP framework to regiment the informational structure of the self at an abstract level. However, at the level of mechanisms, self-structures are embodied. I do not choose between these approaches. I think they (or at least their moderate versions) can be used without any inconsistency. Advocates of both camps agree that FEP is "a deeply unified theory of perception, cognition, and action" (Clark, 2013, p. 186). And they assume that it explains *everything* about the mind and its maximal explanatory scope, and it supports a unified approach to mental functions (Hohwy, 2013, p. 242, 2014, p. 146). I generally submit that FEP owes its great unifying scope to its Bayesian/information-theoretic mathematical framework. However, FEP also describes embodied mechanisms that ground the self-structure. In this sense, the informational structure of the self when stated in terms of FEP represents both abstract models of the self and includes embodied concrete features. I shall unpack this point in the next section.

4.7   A Meta-Theoretical Unifying Informational Framework and Embodied Mechanisms

As I remarked at the end of the previous section, the advocates of FEP praise its great unifying power. However, there are also critics who deny that FEP can explain cognition on the basis of underpinning causal mechanisms that ground cognitive phenomena. Instead, the critics argue, FEP's explanatory pretences are based on its general Bayesian framework. The Bayesian framework is centred on Bayes' theorem, according to which the probability of an event (or hypothesis) given the evidence is equal to the conditional probability of the evidence given the event (or hypothesis) multiplied by the prior probability of the event (or hypothesis) divided by the probability of the evidence. The Bayesian approach to cognition presumes that theories of cognition, perception, action, social cognition, and so on are optimal processes that can be represented in terms of approximate Bayesian models. In this context, Colombo and Series (2012) have argued that theories of Bayesian cognition, in general, can systematise the observational statements about the behaviour of cognisant organisms. They also concede that the Bayesian approach to the study of cognition provides informative predictions about subjects' perceptual performance and the functioning of their neural mechanisms. However, Colombo and colleagues (Colombo & Hartmann, 2015; Colombo & Series, 2012; Colombo & Wright, 2016) also argue that Bayesian inferences cannot be specified in terms of concrete neuronal mechanistic. I think Colombo and colleagues are in right in understanding FEP as a computational framework that represents cognitive dynamics at a high level of abstraction. FEP relies on Bayesian inferences, which form an abstract formal framework, to unify theories of cognition and explain them. It could be argued that it does not explain cognitive phenomena on the basis of concrete neurophysiological mechanisms (Beni, 2018b; Colombo & Wright, 2016).

Let us make a concession regarding Colombo and colleagues' critical insights. It might be that FEP does not explain cognitive phenomena by delving into the study of anatomical or physiological mechanisms. However, even critics must concede that FEP provides a comprehensive and powerful formal framework at a high level of abstraction. Of course, this does not need to mean that FEP cannot explain cognition at all or is devoid of the acclaimed unificatory powers. For one thing, it can be argued that there are plausible unificationist and structuralist theories of explanation that could be used to substantiate the explanatory claims of FEP on the basis of its powerful formal framework instead of its reliance on identification of specific neurophysiological mechanisms (Beni,

2018b; Huneman, 2018; Kitcher, 1989). That being said, for the sake of argument, I can grant that FEP's Bayesian framework does not confer on FEP its great explanatory power. To be more clear, although I think there may be a meaningful relation between the FEP's explanatory power and its formal framework (see Beni, 2018b), I do not defend the existence of such a relation here. Nor do I intend to establish the supremacy of FEP in the face of possible alternatives. My point is that the formal framework of FEP could be used to draw a connection between various theoretical ventures without insisting on the explanatory power of FEP. I simply speak to the widely acknowledged point that FEP provides an inclusive framework for regimenting cognitive mechanisms. More specifically, and with regard to the goal of this book, FEP provides a unifying framework for regimenting the informational structure of the self at a meta-theoretical level. FEP regiments the basic structure of the self at a meta-theoretical level that is characterised in terms of Bayesian inferences and statistical-dynamical relations. It should be noted that although the unifying framework of FEP is forged at a high level of abstraction, FEP relies on this framework to explain some fundamental features of life and cognition globally, despite the fact that free energy "is not a directly measurable physical quantity" (Buckley, Kim, McGregor, & Seth, 2017, p. 57). So, FEP's unificatory framework is not devoid of explanatory power and biological viability. However, I do not here insist on the explanatory power of FEP. What is important from the perspective of SRS is that FEP can draw mathematical relations (in terms of Bayesian relations) between various aspects and functions of selfhood. It provides a nice mathematical representation of the basic structure of the self. Below, I explain why this point is important for SRS.

In Chap. 2, I remarked that some versions of OSR use set/model-theoretic formalism to represent the structure of scientific theories at a meta-theoretical level. In this chapter, I showed that FEP provides a fine Bayesian/information-theoretic framework that can be used to regiment the embodied informational structure of the self at an abstract meta-theoretic level. The embodied informational structures can be specified in terms of their underpinning information-processing mechanisms in CMS and the MNS. FEP, on the other hand, provides a meta-theoretical framework that can represent the abstract informational structure of the self. Such a meta-theoretical representation of the self-structure is useful because it contributes to providing a neat formal understanding of the basic structure of the self as well as the self's representations of the world (i.e., its perceptions) and its phenomenal aspects. I will speak more about an FEP-based account of the phenomenal aspects of the self in the next chapter.

It is also worth mentioning that, although FEP's Bayesian/information-theoretic framework represents the self-structure at an abstract, meta-theoretical level, it is still preferable to the set/model-theoretic framework that is used by the advocates of the orthodox version of OSR (for elaboration, see Beni, 2019b, Chap. 3). This is because, as I have argued in this section, FEP is intimately related to the modelling of psychological phenomena across the fields of cognition, learning, perception, and action. This makes FEP comparatively more adequate framework (than set/model theory) for the purpose of regimenting the informational structure of the self which is a psychological phenomenon. Moreover, as I explained above, FEP can be used to regiment the structure of the self at an abstract representational level. However, there are nice embodied accounts of FEP and predictive coding which underline the importance of embodied, enactive, and ecological aspects of FEP (Clark, 2016a; Gallagher & Allen, 2016; Kiverstein & Rietveld, 2018). This means that FEP's framework is reconcilable to the embodied account of the self. Below, I allude to the enactivist-embodied construal of FEP.

The philosophy of cognitive science includes radical anti-representationalist tendencies, for example, radical embodied and enactivist tendencies (Chemero, 2009; Hutto & Myin, 2013; Varela, Thompson, & Rosch, 1991). These theories inspire a radical embodied-enactivist construal of FEP (Bruineberg et al., 2016; Bruineberg & Rietveld, 2014; Friston, Mattout, & Kilner, 2011; Gallagher & Allen, 2016). The enactivist approach extolls the role of the body and its enacting in the world in the formation of cognition over the brain's capacity for inferring the structure of the world from inside the skull. The embodied-enactivist construal also lays emphasis on the dynamical interconnection between the organism and the environment. It specifies the "agent" and its "environment" as a coupled system and thereby removes the boundary between the organism and its environment. The ecological core of the enactivist approach allows for grounding FEP in the actual mechanisms unveiled by evolutionary psychology and theoretical biology. According to this construal, the organism seeks to change its sensory states by enacting in the world, so as to avoid finding itself in biologically unbearable states. There are no boundaries between the organism and the environment, and active inferences could be understood in terms of the organism's dynamical interplay with the world's windows of affordance, without invoking the representational states endorsed by the inferentialist construal (Gallagher & Allen, 2016; Kiverstein & Rietveld, 2018). Affordances are the opportunities that the environment provides for the organism's action and perception. The development of organisms' finer capacities, for instance, imagination, language, and so on, as well as the survival of all species in general, could be explicated on the basis of the organisms' interface with the affordances (Kiverstein & Rietveld, 2018). Thus, this construal does not leave any room for the notions of "model" and "representation".

In this book, I am advocating a form of moderate embodied theory that does not exclude the notions of "model" and "representation". For example, Burr and Jones argued that taking into account the role of the body, as being realised through *active sensorimotor predictions*, can provide a more comprehensive understanding of FEP. But the emphasis on the role of the body does not contradict Hohwy's account of the inferential mechanisms of representation, albeit if we consider the bodily origins of representations and models (Burr & Jones, 2016). Instead of downplaying the role of either embodied or representational components, the moderate construal seeks to strike a fine balance between two aspects of FEP. In the same way, this construal also underlines the intertwinement of action and perception. The perceptual system is open to the world, and it engages the world that is parsed according to our organism-specific needs and action repertoire rather than representing the mind-independent world through internal models, albeit without going so far to eliminate representations from the picture altogether (Clark, 2016b, p. 195). The moderate embodied construal is compatible with a model-based representationalist account of computation and information processing. This is because, "it is surely that very model-invoking schema that allows us to understand how it is that these looping dynamical regimes arise and enable such spectacular results" (Clark, 2015, pp. 5–6, 2016b, p. 293).

Let me recap. FEP provides a powerful unifying framework for regimenting the self-structure, which is supposed to unify the aspects and elements of the self. However, this does not mean that the self-structure is not in fact embodied. For one thing, I have remarked that the self-structure is grounded in parts of the brain and cognitive system (e.g., CMS, MNS) (I address the issue of the relationship between FEP and CMS theory in the next chapter, Sect. 5.5). Moreover, FEP itself could be construed along the lines of embodiment and enactivism. This allows us for accounting for the unifying power of the infrastructure of the self at the formal level of the Bayesian framework, without ignoring the embodied aspect of the self-structures that are grounded in the organisms and their environment.

4.8   Concluding Remarks

In this chapter, I stated SRS as a viable alternative to the substantivalist theory of the self. The viability of SRS has been substantiated on the basis of its harmony with the results of scientific theories of the field which seem to be diverging rather radically from the substantivalist view. Despite their divergence from substantivalism, scientifically informed theories of the self have not been developed into an alternative metaphysical paradigm. By submitting that the self could be identified in terms of the informational structure that subsumes various aspects of the self as well as various scientific theories that aim to account for these aspects, SRS provides an alternative to substantivalism. SRS's unifying power, that is, its capacity for integrating various aspects of the self on the basis of its informational structure provides further reason for being optimistic about the metaphysical viability of SRS. In the same vein, unlike the substantivalist account of the self, SRS contributes to addressing a form of metaphysical underdetermination of the self, by submitting that philosophical commitments are to be made to the commonality—that is, the common underpinning informational structure—that lies beneath various theoretical conceptions of the self and unifies them. Once the viability of SRS has been established, we must endeavour to specify the informational structure of the self. In this chapter, I specified the structure of the self at two interconnected levels. I specified the embodied informational structure of the self at the level of neurophysiological mechanisms in CMS and the MNS. I also specified the informational structure of the self at a representational, mathematical level, that is, at the level of FEP's formal unifying framework which is mainly constituted by Bayesian inferences. More generally, I follow Piccinini and Bahar (2013) to assume that informational processing in biological cognitive systems or neural systems is sui generis and manifests properties of both digital and analogue forms of information processing. As I have remarked before in this chapter, self-structures are primarily implemented in the brains and cognitive systems, in the sense that they are realised through processing mechanisms performed by brains and cognitive systems. This indicates that self-structures are embodied in the cognitive system. But as enactivist-ecological approaches indicate, the cognitive system is not secluded from its environment. In this sense, informational structures could be extended to latch onto the causal structure of the world. The self-structure can be infused with the structure of environmental information. Please note that the emphasis on the embodied aspects of the self-structure is not in contrast with my remark on the computational-informational nature of the self-structure. As I explained in the previous sections, I am leaning towards a moderate form of embodiment which allows for the existence of informational/computational models of phenomena. To defend this tendency, I refer to some nice attempts at producing computational models of embodied cognition and launching a dialogue between computationalism and embodied cognition (Pezzulo et al., 2011, 2012). So, the structure of the self could be characterised at both abstract and embodied levels. To complete the picture, I need to account for the relationship between FEP and CMS (and their respective coding frameworks). I will do so in the next chapter.

Now that the structural realist core of my conception of selfhood is stated, I will endeavour to show how various phenomenal, social, and ethical aspects of the self could be explicated along the lines of a structural realist theory of the self that is developed in this chapter.

References

Apps, M. A. J., & Tsakiris, M. (2014). The Free-Energy Self: A Predictive Coding Account of Self-Recognition. *Neuroscience & Biobehavioral Reviews, 41*, 85–97. https://doi.org/10.1016/J.NEUBIOREV.2013.01.029

Arnal, L. H., & Giraud, A.-L. (2012). Cortical Oscillations and Sensory Predictions. *Trends in Cognitive Sciences, 16*(7), 390–398. https://doi.org/10.1016/J.TICS.2012.05.003

Beni, M. D. (2016). Structural Realist Account of the Self. *Synthese, 193*(12), 3727–3740. https://doi.org/10.1007/s11229-016-1098-9

Beni, M. D. (2018a). Commentary: The Predictive Processing Paradigm Has Roots in Kant. *Frontiers in Systems Neuroscience, 11*(98). https://doi.org/10.3389/FNSYS.2017.00098

Beni, M. D. (2018b). The Reward of Unification: A Realist Reading of the Predictive Processing Theory. *New Ideas in Psychology, 48*, 21–26. https://doi.org/10.1016/j.newideapsych.2017.10.001

Beni, M. D. (2019a). An Outline of a Unified Theory of the Relational Self: Grounding the Self in the Manifold of Interpersonal Relations. *Phenomenology and the Cognitive Sciences, 18*(3), 473–491. https://doi.org/10.1007/s11097-018-9587-6

Beni, M. D. (2019b). *Cognitive Structural Realism: A Radical Solution to the Problem of Scientific Representation*. Cham: Springer Nature.

Blanke, O., Landis, T., Spinelli, L., & Seeck, M. (2004). Out-of-Body Experience and Autoscopy of Neurological Origin. *Brain, 127*(2), 243–258. https://doi.org/10.1093/brain/awh040

Boyd, R. N. (1980). Scientific Realism and Naturalistic Epistemology. *PSA: Proceedings of the Biennial Meeting of the Philosophy of Science Association*. The University of Chicago Press Philosophy of Science Association. https://doi.org/10.2307/192615

Bruineberg, J., Kiverstein, J., & Rietveld, E. (2016). The Anticipating Brain Is Not a Scientist: The Free-Energy Principle from an Ecological-Enactive Perspective. *Synthese*, 1–28. https://doi.org/10.1007/s11229-016-1239-1

Bruineberg, J., & Rietveld, E. (2014). Self-Organization, Free Energy Minimization, and Optimal Grip on a Field of Affordances. *Frontiers in Human Neuroscience, 8*(599). https://doi.org/10.3389/fnhum.2014.00599

Buckley, C. L., Kim, C. S., McGregor, S., & Seth, A. K. (2017). The Free Energy Principle for Action and Perception: A Mathematical Review. *Journal of Mathematical Psychology, 81*, 55–79. https://doi.org/10.1016/J.JMP.2017.09.004

Burr, C., & Jones, M. (2016). The Body as Laboratory: Prediction-Error Minimization, Embodiment, and Representation. *Philosophical Psychology,* *29*(4), 586–600. https://doi.org/10.1080/09515089.2015.1135238

Chemero, A. (2009). *Radical Embodied Cognitive Science.* Cambridge, MA: MIT Press.

Clark, A. (2013). Whatever Next? Predictive Brains, Situated Agents, and the Future of Cognitive Science. *Behavioral and Brain Sciences, 36*(03), 181–204. https://doi.org/10.1017/S0140525X12000477

Clark, A. (2015). *Predicting Peace: The End of the Representation Wars* (T. Metzinger & J. M. Windt, Eds.). Frankfurt am Main: MIND Group. https://doi.org/10.15502/9783958570979

Clark, A. (2016a). Busting Out: Predictive Brains, Embodied Minds, and the Puzzle of the Evidentiary Veil. *Noûs.* https://doi.org/10.1111/nous.12140

Clark, A. (2016b). *Surfing Uncertainty.* New York: Oxford University Press. https://doi.org/10.1093/acprof:oso/9780190217013.001.0001

Colombo, M., & Hartmann, S. (2015). Bayesian Cognitive Science, Unification, and Explanation. *The British Journal for the Philosophy of Science, 68*(2), axv036. https://doi.org/10.1093/bjps/axv036

Colombo, M., & Series, P. (2012). Bayes in the Brain—On Bayesian Modelling in Neuroscience. *The British Journal for the Philosophy of Science, 63*(3), 697–723. https://doi.org/10.1093/bjps/axr043

Colombo, M., & Wright, C. (2016). Explanatory Pluralism: An Unrewarding Prediction Error for Free Energy Theorists. *Brain and Cognition.* https://doi.org/10.1016/j.bandc.2016.02.003

Esfeld, M. (2009). The Modal Nature of Structures in Ontic Structural Realism. *International Studies in the Philosophy of Science, 23*(2), 179–194. https://doi.org/10.1080/02698590903006917

Floridi, L. (2008). A Defence of Informational Structural Realism. *Synthese, 161*(2), 219–253. https://doi.org/10.1007/s11229-007-9163-z

Fotopoulou, A. (2012). Towards a Psychodynamic Neuroscience. In *From the Couch to the Lab* (pp. 25–46). Oxford University Press. https://doi.org/10.1093/med/9780199600526.003.0003

French, S. (2014). *The Structure of the World: Metaphysics and Representation.* Oxford: Oxford University Press. https://doi.org/10.1093/acprof:oso/9780199684847.001.0001

French, S., & Krause, D. (2010). *Identity in Physics: A Historical, Philosophical, and Formal Analysis.* Oxford University Press. Retrieved from https://global.oup.com/academic/product/identity-in-physics-9780199575633?cc=us&lang=en&

French, S., & Ladyman, J. (2003). Remodelling Structural Realism: Quantum Physics and the Metaphysics of Structure. *Synthese, 136*(1), 31–56. https://doi.org/10.1023/A:1024156116636

Friston, K. J. (2010). The Free-Energy Principle: A Unified Brain Theory? *Nature Reviews Neuroscience, 11*(2), 127–138. https://doi.org/10.1038/nrn2787

Friston, K. J., Mattout, J., & Kilner, J. (2011). Action Understanding and Active Inference. *Biological Cybernetics, 104*(1–2), 137–160. https://doi.org/10.1007/s00422-011-0424-z

Friston, K. J., & Stephan, K. E. (2007). Free-Energy and the Brain. *Synthese, 159*(3), 417–458. https://doi.org/10.1007/s11229-007-9237-y

Gallagher, S. (2013). A Pattern Theory of Self. *Frontiers in Human Neuroscience, 7*, 443. https://doi.org/10.3389/fnhum.2013.00443

Gallagher, S. (2014). Pragmatic Interventions into Enactive and Extended Conceptions of Cognition. *Philosophical Issues, 24*(1), 110–126. https://doi.org/10.1111/phis.12027

Gallagher, S., & Allen, M. (2016). Active Inference, Enactivism and the Hermeneutics of Social Cognition. *Synthese*, 1–22. https://doi.org/10.1007/s11229-016-1269-8

Gallagher, S., & Daly, A. (2018). Dynamical Relations in the Self-Pattern. *Frontiers in Psychology, 9*, 664. https://doi.org/10.3389/fpsyg.2018.00664

Gallese, V. (2003). The Manifold Nature of Interpersonal Relations: The Quest for a Common Mechanism. *Philosophical Transactions of the Royal Society of London B: Biological Sciences, 358*(1431). Retrieved from http://rstb.royalsocietypublishing.org/content/358/1431/517.short

Gallese, V. (2005). Embodied Simulation: From Neurons to Phenomenal Experience. *Phenomenology and the Cognitive Sciences, 4*(1), 23–48. https://doi.org/10.1007/s11097-005-4737-z

Gallese, V. (2014). Bodily Selves in Relation: Embodied Simulation as Second-Person Perspective on Intersubjectivity. *Philosophical Transactions of the Royal Society of London B: Biological Sciences, 369*(1644). Retrieved from http://rstb.royalsocietypublishing.org/content/369/1644/20130177.short

Gallese, V., Eagle, M. N., & Migone, P. (2007). Intentional Attunement: Mirror Neurons and the Neural Underpinnings of Interpersonal Relations. *Journal of the American Psychoanalytic Association, 55*(1), 131–175. https://doi.org/10.1177/00030651070550010601

Gallese, V., Keysers, C., & Rizzolatti, G. (2004). A Unifying View of the Basis of Social Cognition. *Trends in Cognitive Sciences, 8*(9), 396–403. https://doi.org/10.1016/j.tics.2004.07.002

Giambra, L. M. (1995). A Laboratory Method for Investigating Influences on Switching Attention to Task-Unrelated Imagery and Thought. *Consciousness and Cognition, 4*(1), 1–21. https://doi.org/10.1006/ccog.1995.1001

Hohwy, J. (2013). *The Predictive Mind.* Oxford University Press. https://doi.org/10.1093/acprof:oso/9780199682737.001.0001

Hohwy, J. (2014). The Self-Evidencing Brain. *Noûs, 50*(2), 259–285. https://doi.org/10.1111/nous.12062

Hohwy, J., & Paton, B. (2010). Explaining Away the Body: Experiences of Supernaturally Caused Touch and Touch on Non-Hand Objects Within the Rubber Hand Illusion. *PLoS ONE, 5*(2), e9416. https://doi.org/10.1371/journal.pone.0009416

Huang, Y., & Rao, R. P. N. (2011). Predictive Coding. *Wiley Interdisciplinary Reviews: Cognitive Science, 2*(5), 580–593. https://doi.org/10.1002/wcs.142

Huneman, P. (2018). Outlines of a Theory of Structural Explanations. *Philosophical Studies, 175*(3), 665–702. https://doi.org/10.1007/s11098-017-0887-4

Hutto, D. D. (2015). Basic Social Cognition Without Mindreading: Minding Minds Without Attributing Contents. *Synthese, 194*(3), 827–846. https://doi.org/10.1007/s11229-015-0831-0

Hutto, D. D., & Myin, E. (2013). *Radicalizing Enactivism Basic Minds Without Content.* Cambridge, MA: MIT Press.

Kitcher, P. (1989). Explanatory Unification and the Causal Structure of the World. In P. Kitcher & W. Salmon (Eds.), *Scientific Explanation.* Minneapolis, MN: University of Minnesota Press. Retrieved from https://conservancy.umn.edu/handle/11299/185687

Kiverstein, J. D., & Rietveld, E. (2018). Reconceiving Representation-Hungry Cognition: An Ecological-Enactive Proposal. *Adaptive Behavior.* https://doi.org/10.1177/1059712318772778

Ladyman, J. (1998). What Is Structural Realism? *Studies in History and Philosophy of Science Part A, 29*(3), 409–424. https://doi.org/10.1016/S0039-3681(98)80129-5

Ladyman, J. (2007). On the Identity and Diversity of Objects in a Structure. *Proceedings of the Aristotelian Society, Supplementary Volumes.* Oxford University Press The Aristotelian Society. https://doi.org/10.2307/20619100

Ladyman, J., & Ross, D. (2007). *Every Thing Must Go.* Oxford: Oxford University Press. https://doi.org/10.1093/acprof:oso/9780199276196.001.0001

Lenggenhager, B., Tadi, T., Metzinger, T., & Blanke, O. (2007). Video Ergo Sum: Manipulating Bodily Self-Consciousness. *Science, 317*(5841), 1096–1099. https://doi.org/10.1126/science.1143439

Limanowski, J., & Blankenburg, F. (2013). Minimal Self-Models and the Free Energy Principle. *Frontiers in Human Neuroscience, 7*(547). https://doi.org/10.3389/fnhum.2013.00547

Metzinger, T. (2003). *Being No One: The Self-Model Theory of Subjectivity.* Cambridge, MA: MIT Press.

Mitchell, J. P., Macrae, C. N., & Banaji, M. R. (2006). Dissociable Medial Prefrontal Contributions to Judgments of Similar and Dissimilar Others. *Neuron, 50*(4), 655–663. https://doi.org/10.1016/j.neuron.2006.03.040

Northoff, G. (2014a). The Brain's Intrinsic Activity and Inner Time Consciousness in Schizophrenia. *World Psychiatry: Official Journal of the World Psychiatric Association (WPA), 13*(2), 144–145. https://doi.org/10.1002/wps.20122

Northoff, G. (2014b). *Unlocking the Brain: Volume 1: Coding.* New York: Oxford University Press.

Northoff, G., & Bermpohl, F. (2004). Cortical Midline Structures and the Self. *Trends in Cognitive Sciences, 8*(3), 102–107. https://doi.org/10.1016/j.tics.2004.01.004

Northoff, G., Heinzel, A., Northoff, G., Popper, K., Eccles, J., Varela, F., … Churchland, P. (2006). First-Person Neuroscience: A New Methodological Approach for Linking Mental and Neuronal States. *Philosophy, Ethics, and Humanities in Medicine, 1*(1), 3. https://doi.org/10.1186/1747-5341-1-3

Petkova, V. I., & Ehrsson, H. H. (2008). If I Were You: Perceptual Illusion of Body Swapping. *PLoS ONE, 3*(12), e3832. https://doi.org/10.1371/journal.pone.0003832

Pezzulo, G., Barsalou, L. W., Cangelosi, A., Fischer, M. H., McRae, K., & Spivey, M. J. (2011). The Mechanics of Embodiment: A Dialog on Embodiment and Computational Modeling. *Frontiers in Psychology, 2*(5). https://doi.org/10.3389/fpsyg.2011.00005

Pezzulo, G., Barsalou, L. W., Cangelosi, A., Fischer, M. H., McRae, K., & Spivey, M. J. (2012). Computational Grounded Cognition: A New Alliance Between Grounded Cognition and Computational Modeling. *Frontiers in Psychology, 3*, 612. https://doi.org/10.3389/fpsyg.2012.00612

Piccinini, G., & Bahar, S. (2013). Neural Computation and the Computational Theory of Cognition. *Cognitive Science, 37*(3), 453–488. https://doi.org/10.1111/cogs.12012

Psillos, S. (2001). Is Structural Realism Possible? *Philosophy of Science, 68*(S3), S13–S24. https://doi.org/10.1086/392894

Putnam, H. (1975). *Mathematics, Matter, and Method.* Cambridge: Cambridge University Press.

Qin, P., Duncan, N., & Northoff, G. (2013). Why and How Is the Self-Related to the Brain Midline Regions? *Frontiers in Human Neuroscience, 7*(909). https://doi.org/10.3389/fnhum.2013.00909

Qin, P., & Northoff, G. (2011). How Is Our Self Related to Midline Regions and the Default-Mode Network? *NeuroImage, 57*(3), 1221–1233. https://doi.org/10.1016/j.neuroimage.2011.05.028

Rizzolatti, G., & Luppino, G. (2001). The Cortical Motor System. *Neuron, 31*(6), 889–901. Retrieved from http://www.ncbi.nlm.nih.gov/pubmed/11580891

Seth, A. K. (2013). Interoceptive Inference, Emotion, and the Embodied Self. *Trends in Cognitive Sciences, 17*(11), 565–573. https://doi.org/10.1016/J.TICS.2013.09.007

Seth, A. K., Suzuki, K., & Critchley, H. D. (2012). An Interoceptive Predictive Coding Model of Conscious Presence. *Frontiers in Psychology, 2*(395). https://doi.org/10.3389/fpsyg.2011.00395

Seth, A. K., & Tsakiris, M. (2018). Being a Beast Machine: The Somatic Basis of Selfhood. *Trends in Cognitive Sciences, 22*(11), 969–981. https://doi.org/10.1016/J.TICS.2018.08.008

Spaulding, S. (2012). Introduction to Debates on Embodied Social Cognition. *Phenomenology and the Cognitive Sciences, 11*(4), 431–448. https://doi.org/10.1007/s11097-012-9275-x

Uddin, L. Q., Iacoboni, M., Lange, C., & Keenan, J. P. (2007). The Self and Social Cognition: The Role of Cortical Midline Structures and Mirror Neurons. *Trends in Cognitive Sciences, 11*(4), 153–157. https://doi.org/10.1016/j.tics.2007.01.001

Umiltà, M. A., Kohler, E., Gallese, V., Fogassi, L., Fadiga, L., Keysers, C., & Rizzolatti, G. (2001). I Know What You Are Doing. *Neuron, 31*(1), 155–165. https://doi.org/10.1016/S0896-6273(01)00337-3

Varela, F. J., Thompson, E., & Rosch, E. (1991). *The Embodied Mind: Cognitive Science and Human Experience.* Cambridge, MA: MIT Press.

Weiler, M., Northoff, G., Damasceno, B. P., & Balthazar, M. L. F. (2016). Self, Cortical Midline Structures and the Resting State: Implications for Alzheimer's Disease. *Neuroscience & Biobehavioral Reviews, 68*, 245–255. https://doi.org/10.1016/j.neubiorev.2016.05.028

Williford, K., Bennequin, D., Friston, K., & Rudrauf, D. (2018). The Projective Consciousness Model and Phenomenal Selfhood. *Frontiers in Psychology, 9*(2571). https://doi.org/10.3389/fpsyg.2018.02571

Wohlschläger, A., Haggard, P., Gesierich, B., & Prinz, W. (2003). The Perceived Onset Time of Self- and Other-Generated Actions. *Psychological Science, 14*(6), 586–591. https://doi.org/10.1046/j.0956-7976.2003.psci_1469.x

5

Phenomenal Aspects of the Self

5.1 Structural Theories of Consciousness

The self has some phenomenal aspects. That is to say, the self is capable of having subjective experiences, and it accommodates conscious and intentional states. In this chapter, I will show how a structural realist account of the self could be supplemented with a structural realist account of the phenomenal aspects of the self, that is, with a structural realist theory of intentionality as well as a structural realist theory of consciousness. Literature includes quite a few theories of consciousness, only three of which will be surveyed in this chapter. In other words, the chapter aims to review three specific theories of consciousness that are in line with various aspects of the structural realist theory of the self (SRS; as being presented in the previous chapter). These are the Integrated Information Theory of consciousness (IIT), the resting-state-based theory of consciousness, and the Free Energy Principle (FEP)-based theory of (self-) consciousness. As I will argue in this chapter, all three of these theories

Parts of this chapter are reprinted with the kind permission from Springer Nature, and The *Journal of Consciousness Studies* (Ingenta Connect Publication).

© The Author(s) 2019
M. D. Beni, *Structuring the Self*, New Directions in Philosophy and Cognitive Science,
https://doi.org/10.1007/978-3-030-31102-5_5

could be construed along the lines of SR. This is an intriguing result; there are several (at least three) structural realist theories of consciousness. Which one of them could be adopted as the right structural realist of consciousness? In this chapter, I will argue that a structural realist does not need to choose one of these theories as the exclusively right theory of consciousness. Three theories of consciousness that are discussed in this chapter account for conscious phenomena in various ways and at different explanatory levels. IIT, for example, provides a mathematical characterisation of the experiential aspects of the self, whereas the resting-state-based account aims to explicate consciousness on the basis of resting state activity of the default mode neural network and difference-based coding. FEP is a theory of computational neuroscience and theoretical biology that underpins yet another definition of self-conscious phenomena on the basis of a variational Bayesian formulation of epistemic functions of the kinds of organisms that can form temporally deep models of their own future. We do not need to choose between these theories. It can be easily assumed that these theories describe various parts of the basic structure of phenomenal aspects of the self at various levels. That is to say, all of the theories of phenomenal aspects of the self that I will present in this chapter are in harmony with SR. Accordingly, I use a structuralist strategy to show how the phenomenal aspect of the self could be specified in terms of the common underlying structure or the commonality between the various theories that I introduce in this chapter.

Before proceeding further, I shall remark that the three mentioned accounts are not the only plausible neuroscientific theories of consciousness. There are other viable theories, such as the global workplace model (Dehaene, Kerszberg, & Changeux, 2006), the recurrent processing theory (Lamme, 2006), and the higher-order theory of consciousness (Rosenthal, 2012). There may be links between the absent theories and theories that are presented in this chapter. For example, FEP-based account resembles higher-order theory (because Friston, too, assumes that consciousness requires self-consciousness). Also, Northoff et al. argue that it is possible to draw fine category-theoretical relations between some of the theories that are not presented here and IIT (Northoff, Tsuchiya, & Saigo, 2019). However, there is not enough space to cover all of the theories of consciousness in this chapter. Moreover, I have already provided a neuroscientific account of the basic structure of the self in terms of the resting state activity and FEP. So, it seems natural to build a structural account of phenomenal aspects upon what I have already said about the basic structure of the self.

Moreover, please note that I am advocating a non-eliminativist version of SRS (NSR). This means that some aspects and elements that are traditionally associated with the self, for example, sense of agency, ownership, and so on, can still feature in my structural account of phenomenal aspects of the self as non-structural elements. Actually, it seems that the accounts that I outline in this chapter, for instance, FEP-based account, can explain some non-structural elements of the self, for example, agency viably enough (for explanation, see Friston, Samothrakis, & Montague, 2012; Limanowski & Friston, 2018).

Be that as may, in Sects. 5.2, 5.3, and 5.4, I outline the three mentioned theories that describe different parts of the structure of consciousness at various levels. Section 5.5 is especially important because it identifies the structural relations that relate the three mentioned theories of consciousness together. The existence of structural relations between these theories indicates that consciousness has a basic unified structure. To substantiate this point, in this chapter I will argue that the FEP-based theory is structurally connected with the resting state activity because it presupposes difference-based coding. Because, as I will argue, the FEP-based theory is also related to IIT, the three theories together describe the basic informational structure of consciousness. The existence of various scientifically respectful theories of consciousness may tend to cause a state of underdetermination of the theory of consciousness by diverse scientific accounts. A structural realist strategy would help us to constrain the diversity of the scientific theories of consciousness by suggesting that these theories are parts of a unifying structure which represent the basic infrastructure of consciousness.

Intentionality, or the capacity of mental states for being about something, is another phenomenal aspect of the self that will be examined in this chapter. If we take intentionality as a serious property of the mind (or the self) at all, we have to face the problem of intentional inexistence. In this chapter, I also show how it is that a structural realist theory of intentionality can provide a handle on the problem of intentional inexistence.

Thus, the chapter unfolds structural realist accounts of some of the most important phenomenal aspects of the self, for example, consciousness, self-consciousness, and intentionality.

5.2   Integrated Information Theory

5.2.1   IIT

This chapter aims to account for the phenomenal aspects of the self, that is, the self's capacity for having subjective experience and consciousness, along the lines of SR. The first theory that has a propensity for accomplishing this task, that is, setting the example of a structural realist theory of consciousness, is the Integrated Information Theory of consciousness (IIT for short). This theory has been developed by Giulio Tononi and Christoph Koch, amongst others (Joshi et al., 2013; Koch & Tononi, 2013; Oizumi et al., 2014; Tononi, 2004, 2015; Tononi, Boly, Massimini, & Koch, 2016; Tononi & Koch, 2015). IIT aims to account for the experiential features of consciousness in terms of the properties of an integrated information system. The integrated information system consists of causally interacting parts (Koch, 2012). Various elements of conscious experience as the constituting components of the informational system are integrated together on the basis of the underpinning informational structure. The informational structure is grounded in a physical causal structure. And the causal structure of consciousness subsumes various mechanisms that altogether realise the phenomenal aspects of the self. In this section, I show how IIT characterises experiential features of agency and consciousness mathematically and grounds them in the underpinning causal structure. The causal structure itself can be specified in terms of mechanisms of release of the synaptic vesicle, action potential, and calcium ions, as well as the embodied mechanisms of the brain's cortical structure, and so on. In this context, IIT introduces the notion of integrated information represented by (Φ). Φ denotes the size of the conscious repertoire that latches onto a network of causally interacting components of the system. It provides a quantitative measure of the irreducibility of a cause-effect structure of a unidirectional partition. The outline of IIT that I present here does not elaborate on the fine mathematical details that bestow upon IIT its exquisite technical plausibility. Such details can be found in numerous works of Tononi and colleagues (Op. Cit.). My point is that IIT's scientifically informed theory of consciousness is in line with an informational version of SR (ISR), which has been outlined in Chap. 2. The structural realist theory of the self, too, has been unfolded in terms of ISR (see Chaps. 3 and 4). And the tendency to explicate consciousness on the basis of informational structures indicates that IIT is in complete agreement with the structural realist theory of the self which also aims to specify selfhood and its phenomenal aspects on the basis of informational structures. Below, I shall draw on Oizumi et al. (2014) to briefly outline IIT.

IIT comes with a neat mathematical formalism, based on the axiomatic formulation of the essential properties of experience. Axioms of IIT that capture the essence of consciousness are stated in the following order (based on Tononi, 2015):

- Intrinsic existence: consciousness exist. Each experience actually exists too, and it exists from its own intrinsic perspective, instead of the perspective of external observers.
- Composition: consciousness is structured, and the structure of consciousness can be characterised on the basis of multiple phenomenological distinctions.
- Information: consciousness is specific and each experience is different from other sets of specific phenomenal experiences.
- Integration: consciousness is unified and experiences are irreducible to disintegrated subsets of phenomenal distinctions.
- Exclusion: consciousness is definite and each experience has its own characteristics and flows in its own pace.

These axioms characterise the main features of subjective experience. IIT's formalism also includes postulates which specify the properties of the physical substrate of consciousness in accordance with the axiomatised properties of consciousness. These are postulates of intrinsic existence, composition, information, integration, and exclusion. In this fashion, IIT mathematically characterises both phenomenological properties of experience (in terms of axioms) and the informational/causal properties of the physical systems (in terms of postulates). This definition of consciousness is quantified, and it draws connections between levels of consciousness of a physical system on the one hand and the repertoire of the causal informational structures that are available to the system as a whole on the other. To be more specific, postulates of IIT specify the quality of consciousness in terms of a *maximally irreducible conceptual structure* (MICS) generated by a complex of elements. IIT specifies the quantity of consciousness in terms of maximally integrated conceptual information denoted by Φ^{MAX}. In each conscious experience, MICS identifies the thing to which the concept refers, whereas Φ^{MAX} provides a quantitative measure for the concepts that are present. That is to say, at the level of individual mechanisms, IIT defines concept φ^{MAX} as a mechanism that specifies a maximally irreducible cause-effect repertoire in the narrow sense. At the level of the whole system of mechanisms, IIT characterises complex Φ^{MAX} as a set of elements whose mechanisms specify a maximally irreducible conceptual structure in the broad sense.

IIT specifies consciousness in terms of information theory. I shall proceed to provide a clear conception of IIT's information-theoretic framework immediately. Tononi's (2004) has used Shannon's classical theory of information to unfold an early version of IIT. Shannon's theory characterises the quantity of uncertainty or "disorder" $H(X)$ in a collection of messages in terms of a probability distribution P over the set of messages (Shannon & Weaver, 1949). Accordingly, the communication entropy of X for a set of messages $x_i = (I = 1, \ldots n)$ can be calculated in the following way:

$$H(X) = -\sum_{i=1,n} P(x_i) \log P(x_i)$$

More recently, some advocates of IIT (e.g., Oizumi et al., 2014) have stated their conception of intrinsic information differently from what is indicated by Shannon's theory. According to this new statement, information can be identified in terms of how a mechanism in its current state constrains the system's potential past and future states. Of course, this definition is not overwhelmingly different to Shannon's classical notion, because, as Oiziumi et al. submit, Shannon's information, too, identifies in terms of the relation between the past states and current states of the same system. Despite that, Oiziumi et al. (Op. Cit.) argue that the main difference between this new conception and Shannon's notion of information is that the latter can be always specified from the extrinsic perspective of an observer who assesses the statistical dependency, whereas IIT's notion of information identifies from the intrinsic perspective of a system in terms of the differences that make a difference to it (see Oizumi et al., 2014, Text 3). The information-theoretic nature of IIT's conception of consciousness makes it prone to be construed along the lines of SR, and thereby it can be compatible with the structural realist theory of the self (SRS). There are, however, further details that provide a more solid purchase for construing IIT along the lines of SR. I have previously fleshed out grounds for a structuralist reading of IIT in response to John Searle's attack on IIT (Beni, 2018a). In the next section, I revisit these grounds.

5.2.2 The Structuralist Tendency of IIT

In his critique of IIT, Searle has built upon the syntactical/semantical dichotomy to argue that IIT's notion of information as measurable mathematical quantity is a strictly syntactical notion (Searle, 2013). Searle presupposes that a syntactical theory fails to account for consciousness which is allegedly contentful, and he concludes that IIT's endeavour for explicating consciousness as some measurable mathematical quantity is pointless. Searle's reservation about reducing the content to a formal account of consciousness is understandable. However, given a structuralist construal of IIT, IIT does not aim to reduce the content to the mathematical relations so much as to submit that the content could be identified in virtue of structural relations. I shall unpack this last remark immediately.

Searle argues that IIT, or any other formal account for that matter, cannot reduce the content of consciousness. However, IIT, especially when construed along the lines of structuralism, does not aim to reduce the content of consciousness at all. Instead of attempting the reduction of the content, IIT evinces that the content of consciousness could be identified in terms of the maximally irreducible conceptual structure (MICS) which is the local maximum of integrated conceptual information, that is, Φ^{MAX}. This point is expressed by the central identity principle of IIT, which aims to establish an identity relationship between certain kinds of structures (conceptual structures) and the quality of experience. It will be still possible to contend that the informational structures could represent the richness of the subjective consciousness experience if one takes a traditional stance on the nature of conscious experiences. For example, as Searle (2013) has remarked, the mathematical theory of information cannot account for the contentfulness of subjective aspects of the experience. This criticism presumes that content cannot be specified in terms of (or represented by) mathematical, informational structures. But there is no reason to assume that the content of consciousness needs to be a well-defined (mental) *individual object*, that is, an object that can be identified in terms suggested by the orthodox object-oriented metaphysics, nor has Searle provided an argument to the effect. Searle's criticism is loyal to the object-oriented metaphysics. However, it is possible to entertain a structuralist conception of subjective experience, according to which the content could be specified in virtue of its place in the conceptual or causal structures of IIT. In support of this structuralist construal, it should be noted that there already exist a number of flourishing structuralist theories of mental representation (Cummins, 1996; O'Brien & Opie, 2004; Shagrir, 2012; Shea, 2016; Swoyer, 1991). And in the previous chapters of this book, I advocated a structuralist theory of the selfhood. In harmony with this structuralist tendency, it is possible to defend a structuralist account of consciousness.

Although this structuralist approach defies the object-oriented intuitions, it comes with a neat formal formulation and is in harmony with recent scientific accounts of consciousness. Here, speaking of scientific findings does not mean that IIT is exclusively based on neurological theories of the brain. Rather it is presented as a neat scientific formulation of the nature of conscious experiences. The characterisation of conscious experience as a conceptual structure specified by a complex of units allows for a structuralist construal. The conceptual structure of experiences is irreducible to non-interdependent components (Tononi, 2017b, p. 244), and it is defined as a maximally irreducible cause-effect structure composed of concepts and their relations. The principle of central identity which has been mentioned earlier in this chapter indicates that every experience is identical with a conceptual structure that is maximally and intrinsically irreducible (ibid., p. 248). All of these add up to the conclusion that IIT is compatible with a structuralist construal. The point, about the structuralist tendency of IIT, has been somewhat acknowledged by the advocates of IIT. Tononi, for example, has submitted that "the structure of experience and its meaning can only be reflected by the composition of concepts within a conceptual structure, and not by neural states, which are 'flat' rather than structured" (Tononi, 2017b, p. 250). In the same vein, the axiom of composition indicates that consciousness itself is structured and each experience has an internal structure (Tononi, 2017b, p. 244). And finally, IIT's conception of information as "differences that make a difference" is completely in line with a structuralist tendency. This definition represents information as a totally relational, difference-based entity, in contrast to an individual entity. It characterises the informational objects structurally and on the basis of difference, rather than their intrinsic and individual properties. Although the informational structure of consciousness is capable of subsuming and unifying diverse experiences (i.e., contents) by underlying them, it cannot be reduced to the individual content of specific experiences that feature in the structure. This characterisation of the informational structure of consciousness conforms to the philosophical core of SR according to which the structure does not reduce to the individual objects. In other words, the structural realist thesis can provide a basis for interpreting IIT, which, among other things, indicates that the integration postulate applies to *relations* among cause-effect repertoires by holding that if two or more concepts overlap over some units, they are *irreducibly* related (Tononi, 2017b, p. 246). This means that the causal power of the structured concepts combines them into a unified whole. The concepts cannot be *disentangled* into independent cause-effect repertoires. Relations among the parts exist only if the parts cannot be disentangled within the whole, and consciousness can exist only in terms of that unified whole (Tononi, 2017a, p. 622). This provides a relational account of integration that could be defined in precise quantified terms. In the same way, the postulate of exclusion applies to relations among concepts (see Tononi, 2017b).

Let us recap. In this section, I strived to show that IIT's reliance on the role of the underpinning structure of consciousness enables us to construe it as a structuralist theory of consciousness. It provides a structuralist account of the content, instead of accounting for the content of consciousness along the lines of the orthodox metaphysics of subjective experiences. The content of consciousness can be identified mainly by virtue of the constellation of conceptual substructures that underpin experiences. Given that consciousness needs to be grounded in the physical substrate, IIT also provides a structuralist account of the physical underpinning of consciousness. Tononi defines the physical substrate of consciousness in terms of a Q-fold (a composition of concepts bound by relations) within the conceptual structure or quale (Tononi, 2017a, p. 625). And as concepts and meanings are supposed to be the same, it could be assumed that whenever there is an overlap of the purviews of different concepts, their meanings are not independent, but *bound* together (are *composed*) structurally (ibid., p. 628). Thus concepts that overlap are not independent, and they impose joint constraints on shared units. In the same vein, Barrett shows how to ground information in the physical substrate, namely by arguing that consciousness could be an intrinsic property of the brain's electromagnetic fields with rich spatiotemporal structures (Barrett, 2014). This is because, according to this proposal, only electromagnetic fields uniquely lend themselves to the formation of complex structures that are at issue in IIT. The proposal focuses on the physical substrate of IIT with an emphasis on the role of the rich spatiotemporal structures which realise IIT's conceptual structures.

The role of the physical structures that embody the informational structure of consciousness is rather important from the point of view of this book. As I have argued in the previous chapter, ontologically, the underpinning structure of the self must be specified in terms of embodied informational structures. IIT characterises the experiential aspects of the self in terms of abstract formalism, that is, abstract informational structures. However, it finally grounds the abstract informational structures in physical or embodied informational structures. Given IIT's account of the role of the physical substrate, it could be assumed that informational structures of consciousness are grounded in the mechanisms which accommodate cause-effect power. That is to say, embodied or grounded informational structures harbour irreducible cause-effect power. Such embodied informational structures can be a subject of ontological commitments. A look at IIT's postulates provides further reasons to appreciate the grounded nature of the causal structures. For instance, just as the composition axiom submits that consciousness is structured, the zeroth postulate of IIT indicates that *existence is cause-effect power*, and the first postulate (intrinsicality) asserts that *the physical substrate of consciousness must have cause-effect power for itself*. Thus, the structure of IIT is causal and grounded. Again, this is in complete agreement with SRS which, as I explained in the previous chapter, specifies the underpinning patterns of selfhood in terms of embodied informational structures.

Be that as it may, the conceptual structure determines the quantity and quality of the experience. IIT's central identity principle holds that the experience is *identical* with a *conceptual structure*. The conceptual structure is maximally irreducible to the components that feature in it. And the account of the physical base of consciousness can be unfolded in terms of a structuralist account of electromagnetic fields. On such grounds, it could be argued that IIT's account of the experiential feature of consciousness as well as its take on the physical substrate of consciousness is in harmony with a structural realist theory of consciousness. In the same way, it is worth mentioning that the experimental studies that confirm the empirical plausibility of IIT support the structuralist construal of consciousness. For example, studies based on the use of cranial magnetic stimulation (TMS) and high-density electroencephalography (EEG) techniques suggest that loss and recovery of consciousness (in sleep, anaesthesia, or brain lesions) corresponds to the breakthrough or restoration of the conceptual *structures* of the brain's information integration (Tononi & Koch, 2015). This indicates that what is responsible for the conscious experience is the enduring *informational structure* of the consciousness, rather than any specific neuronal or biological mechanisms and processes. All of these add up to the conclusion that IIT is completely consistent with SR in general and SRS in particular

Although I construed IIT as a modest theory that only aims to specify the basic structure of consciousness, some scepticism about the philosophical credentials of IIT may endure. For example, it might be objected that there is no concrete reason for accepting IIT's axioms or for thinking that phenomenal consciousness/subjective experience can be cashed out in terms of maximally integrated conceptual information. Some of these objections are quite general and may target any form of structuralist account of mental phenomena. For example, it might seem implausible to suppose that there is an identity relationship between certain kinds of structures (conceptual structures) and the quality of experience. However, in this book, I do not defend the identity of the informational structures with the self or its phenomenal aspects. I simply submit that these structures constitute the basic structure of the self that could underpin the richness of experience and include phenomenal elements such as sense of agency. However, there are also more specific objections that have their root in the fact IIT is mainly concerned with characterising features of subjective experience on the basis of a purely formal characterisation of informational structures (say, instead of the neurophysiological substrate of consciousness). In response to such objections, I will proceed to support the structural account of phenomenal aspects of the self with other theories that are not liable to this specific objection (I also account for the relation between IIT and other theories in Sect. 5.5). Georg Northoff offered a neurophysiological theory of consciousness that at least at first glance is mainly concerned with the neurological information-processing mechanisms and coding systems that realise the underpinning structure of consciousness. I shall overview Northoff's theory in the next section.

5.3   Resting State Theory of the Self

5.3.1   A Neurophysiological Account of Selfhood

In Chap. 4, I suggested that it is possible to specify the underlying structure of the self in terms of neuronal patterns implemented in the cortical midline structures (CMS) and the Default Mode Network (DMN). The characterisation of the self-structure was taking place at the neurophysiological level. It is also possible to provide a structuralist theory of phenomenal aspects of the self, such as the capacity for being conscious or having subjective experience, in neurophysiological terms, that is, terms of coding processes at the neurological level in CMS. This neurophysiological account is supported by scientific facts about the structure of the brain and its intrinsic (i.e., resting state) activity in DMN. As I have explained in the previous chapter, CMS is a component of the brain's DMN which represents the activity of the resting state of the brain. The CMS theory of the self, as being defended by Northoff, accounts for the self-basic structure on the basis of the resting state activity which is embodied in CMS. Also, the brain's special way of coding in CMS and DMN plays a part in the emergence of consciousness. According to Northoff, difference-based coding, which underpins the resting state activity, is the encoding strategy that the brain uses to represent the structure of the self and its phenomenal properties (Northoff, 2014b). The resting state activity realises patterns of spontaneous connectivity between various brain regions. Such connections are observable when the brain is not affected by external stimuli or tasks (Greicius, Krasnow, Reiss, & Menon, 2003). This means that the brain's resting state activity can be specified statistically in terms of the functional connectivity, which forms the class of connections between various brain regions that engage in diverse activities across different spatial points, on the one hand, and low-frequency fluctuation (between 0.001 Hz and 4 Hz) across discrete temporal points on the other. Thus, the difference between diverse spatial and temporal points in (the cortical layers of) the brain generates the spatiotemporal structure that underlies the brain's resting state activity. This approach to the self (and consciousness) is not based on modelling the behaviour of the neuronal population when the brain is engaged in processing stimuli as external entities that elicit the brain's functions from the outside. Instead, Northoff offers to account for the self and its phenomenal aspects by elaborating on the activity of the brain's Default Mode Network and resting state activity, which amounts to a form of intrinsic activity generated spontaneously by the brain in the absence of the extrinsic factor, that is, the stimulus (hence the term "intrinsic activity") (also see Northoff, 2018). As I argued in the previous chapter, according to Northoff, the intrinsic activity of the brain forms physical structures which (when formed at CMS) can embody the self. In the same vein, Northoff argues that the spatiotemporal structure of the brain and its resting state activity is predisposed to embody the organisation of the phenomenal aspects of the self (Northoff, 2014b, p. xix). Northoff calls this the resting-state-based theory of consciousness. According to him,

> The "resting-state-based account of consciousness" claims that the brain's intrinsic activity predisposes and thus makes necessary and unavoidable the possible association of its otherwise purely neuronal resting state and stimulus-induced activity with consciousness and its phenomenal features. (Northoff, 2014b, p. xix)

In a nutshell, Northoff assumes that the brain's Default Mode Network enables it to react to the stimulus (which is an external element) in certain ways and thereby the resting state activity predisposes the correlation between consciousness (as a phenomenal aspect of the self) and neuronal activity of the brain, extrinsic activity and reaction to external stimuli included. The brain's intrinsic features, defined in terms of resting state activity, enable the brain to react to the stimuli from the body and stimuli from the environment, and it predisposes the association between the neuronal processing of external and bodily stimuli with conscious states. Below, I explain that this account of consciousness is in harmony with the structuralist agenda of this book.

5.3.2   Difference-Based Coding and Structuralism About the Self

To generate the spatiotemporal structure of the Default Mode Network, the sensory cortex of the brain encodes the statistical frequency distribution of the extrinsic stimuli on the basis of encoding differences between discrete points in space and time. This provides a plausible and economical way of modelling the environment, because as Northoff explains the relationship between stimuli (as differences between discrete points in space and time) and neurons or brain regions that encode the stimuli is not one-to-one but many-to-one (see his account of sparse coding in

Northoff (2014a), part III). A small number of neurons can encode the differences between a large amount of discrete spatiotemporal points (i.e., a proportionally large amount of information), and therefore the difference-based coding is economical. According to Northoff, the spatiotemporal structure of the Default Mode Network, which underlies the resting state activity, is generated by this economical type of coding, that is, the brain's difference-based coding. Here, the general idea is that "the spatiotemporal structure is based on the encoding of temporal and spatial differences between different stimuli rather than on the stimuli themselves and their respective physical features" (Northoff, 2014b, p. xxx).

The resting state theory can be construed as a structuralist account of consciousness. This is because the brain's Default Mode Network (which embodies consciousness) is generated by coding the relations between sets of stimuli, rather than coding the properties of the stimuli or relata. Because the account gives priority to relations over individual objects, it comes with a clear structuralist tendency. The structure of the resting state's spatiotemporal structure consists of statistically forged patterns. These patterns are forged through "encoding of the statistical frequency distribution of the stimuli across different discrete points in physical time and space, that is, the natural statistics of the encoded stimuli" (Northoff, 2014b, p. xxxi). Modelling the resting state activity structurally, in terms of difference-based coding instead of coding of stimuli (as relata), contributes to understanding Northoff's theory of consciousness as a structuralist theory. To be more precise, the alignment between neural coding—that is, coding of stimuli via capturing differences between spatiotemporal points—and the brain resting state (or intrinsic) activity—as the brain's virtual spatiotemporal structure—predisposes the constitution of consciousness. Northoff has provided a detailed account of the neuroscience of interaction between the resting state activity and stimulus-based coding (Northoff, 2014a, part IV). He submits that the spatiotemporal structure of the resting state activity, forged statistically through difference-based coding, provides the form of consciousness. The intrinsic structure of the brain integrates, structures, and organises various aspects (or contents) of consciousness. Contents of consciousness could be identified in virtue of their status in the spatiotemporal grid that is imposed by the resting state activity. Thus the resting state theory of consciousness is in line with SRS and its structuralist tendency.

Let me recap. In the previous chapter, I argued that it is possible to specify the structure of the self at the neurophysiological level on the basis of the information processing in CMS. CMS is part of the Default Mode Network that realises the resting state activity. And Northoff's account of consciousness presumes that the brain's resting state activity predisposes the constitution of consciousness. Thus, this theory of consciousness is in line with SRS, especially the part of the theory that has been inspired by Northoff's CMS theory of the self. In the same vein, owing to its reliance on difference-based coding—that is, coding of differences between discrete points in space and time—Northoff's account of consciousness comes with a clear structuralist tendency. This is in harmony with the main insight of this book, which supports a structural realist theory of the self and its features. That being said, Northoff's neural hypothesis about consciousness highlights the predisposition of the resting state activity (aligned with the difference-based coding) for the constitution of consciousness. But is "predisposition relation" strong enough to provide an informative account of consciousness? Below, I shall consider this question.

To clarify his stance, Northoff argues that "consciousness and its phenomenal features are supposed to show a spatiotemporal structure that can be traced back to the one of the resting state and its alignment to the environment" (Northoff, 2014b, p. xxxi). However, in my view, the problem is that the capacity for *being predisposed* by a specific kind of neuronal activity and *being traceable* to it are not quite sufficient for providing a clear understanding of the nature of consciousness. It is true that Northoff struggles (heroically in my opinion) to develop his neural hypothesis of consciousness to a neurophenomenal hypothesis which aims to account for the intrinsic features of consciousness, for instance, its conceptual and phenomenal features. However, although the theory may specify the correlations between some neuronal populations and some phenomenal content of consciousness—for example, the phenomenal content in dreams (Northoff, 2014b, chapter 26)—it cannot explain the phenomenal experience of the content of consciousness on the basis of the neuronal underpinning mechanisms. One reason for this scepticism is that "correlations" and "predispositions" are not reliable theoretical terms, and they are not strong enough to explain consciousness viably.

In agreement with many other philosophers, I assume that explanations are viable when they are based on facts about patterns of causal efficiency and causal powers (Psillos, 2002; Salmon, 1984). The resting-state-based theory of the self does not include a causal explanation of consciousness, and Northoff does not aim to advocate a causal theory of explanation. Instead, he seeks to reconcile his frequent use of resting state activity as what "predisposes" consciousness to philosophical discourse on "dispositions" (Northoff, 2014b, p. lvii ff.). According to Northoff, the resting state activity predisposes (possible) consciousness in the same way that irregular alignment of crystal—identified with the disposition to be fragile—predisposes possible shattering rather than actual shattering. The fragile object, as being predisposed by the irregular alignment of crystal, would actually shatter only when imposed to an external force. In the same way, the actual consciousness, as being predisposed by resting state activity, emerges in the presence of external stimuli (Northoff, 2014b, p. lvii ff.). According to this approach, the resting state activity and the neurophysiological regions that embody it are just making consciousness possible, instead of providing necessary and sufficient condition for its generation. The problem with this attempt at specifying the neural underpinning of consciousness is that it relies too much on dispositions (and correlations). But this discourse does not ipso facto provide viable explanations of consciousness in terms of patterns of causal relations between underpinning neuronal mechanisms and phenomenal aspects of the self. The issue of insufficiency of appealing to dispositions to explicate the phenomena has been already raised in the philosophy of science (Armstrong, 1968; Mackie, 1973, 1977). And in the face of expressed scepticism about the philosophical credential of the dispositions discourse, Northoff's reliance on terms such as "disposition" and "correlation" may seem questionable. Let me elaborate.

Dispositions as such are not sufficient for underpinning viable explanations. Accounting for the fact that one is drinking water by saying that one is thirsty (assuming that being thirsty is a disposition) does not explain the phenomenon viably; at least not in the sense that explanations based on non-dispositional fundamental physiological and biological facts (say, about the mechanisms of metabolism) can furnish viable explanations. As Armstrong argues, explaining phenomena on the basis of dispositions would be unsuccessful (and would lean towards operationalism) unless we could show that "the disposition is contingently associated with a certain non-dispositional state, presumably a state of the brain" (Armstrong, 1968, p. 59). To prevent this eventuality, Armstrong suggests that it is best to think of such non-dispositional states as causes or causal factors (Armstrong, 1968, p. 88). This is because, without being grounded in non-dispositional mechanisms that can be identified as causal factors that underlie informative viable explanations, the disposition-based discourse would be devoid of explanatory value. It is true that hypotheses about the resting state activity and difference-based coding may be understood as non-dispositional facts that support Northoff's account. However, Northoff's elaboration on the role of the resting state activity and difference-based coding as neural mechanisms that predispose consciousness does not identify the causal patterns that ground consciousness in the resting state activity. To be clear, I think the resting-state-based theory of consciousness is interesting and in harmony with the structuralist tendency of this book. However, I think Northoff's reliance on predispositions and correlations does not contribute to a robust realist philosophical discourse on consciousness. Additionally, although Northoff supports his theory by exquisite neurological details based on imaging experiments that model neurological mechanisms of information processing in the brain, his account of the difference-based coding is too general. That is to say, as Northoff acknowledges, the difference-based coding is the common currency of the brain's neural activity, and it provides the basis for accounting for all kinds of neuronal activities. In this sense, it does not provide a specific model of consciousness in the sense that a more detailed theoretical model could. This does not mean of course to question the plausibility of the resting-state-based account of consciousness as a theory concerned with the most fundamental and general kind of neural activity that makes consciousness possible. I simply submit that there are features of consciousness that are not fathomed by Northoff's theory. In the next section, I outline yet another structuralist account of consciousness that seeks to offer more detailed models of consciousness and self-consciousness by invoking the criteria of temporal depth and counterfactual depth of variational Bayesian models.

5.4   Free Energy and (Self)Consciousness

In the previous chapter, I pointed out that predictive processing and the Free Energy Principle (FEP) provide a meta-theoretical framework for regimenting the informational structure of the self. Recently, Friston has presented the FEP-based account of consciousness. In this section, I will offer a critical analysis of the FEP-based theory of consciousness.

5.4.1   The Free Energy Principle, Some Platitudes

I have briefly introduced FEP in the previous chapter. Here, I add some details. As I remarked, FEP underlies a unifying theory of cognition, perception, action, and learning. Karl Friston articulated the theory on a number of occasions and applied it to global accounts of embodied perception and cognition (Adams, Shipp, & Friston, 2013; Friston, 2010; Friston & Kiebel, 2009). FEP lines up with the second law of thermodynamics and Gibbs' notion of free energy, as well as with the classical theory of information.

Free energy can be characterised as (Gibbs) energy minus the entropy of the system, where the energies and entropies pertain not to thermodynamics but to probability distributions encoded by physical states (Friston et al., 2015, p. 191). Given the second law of thermodynamics, the total entropy of natural systems tends to increase. However, some systems are capable of self-organisation and re-organisation, and they tend to resist the dispersing effect of their environment. When articulated in terms of FEP, this means that in order to maximise their survival, organisms have to minimise their free energy. The theory is universal in the sense that it models the activity of all kinds of self-organising systems from viruses to ecosystems and evolution. Thus, FEP has been characterised in terms of variational Bayesianism (Friston et al., 2012; Ramstead,

Badcock, & Friston, 2018). Free energy itself is an information-theoretic measure. According to Friston (2010, p. 2), free energy can be characterised as:

$$F = -\Big\langle \ln\ p\big(s \sim ,\mid \vartheta, \mid m\big)\Big\rangle q + \Big\langle \ln\ q\big(\vartheta\mid\mu\big)\Big\rangle q$$

Here, $s\sim$ is the set of sensory signals and their motions, $p(s\sim, \vartheta\mid m)$ is the probability density that generates sensory samples and their causes, and $q(\vartheta\mid\mu)$ is the recognition density (i.e., the organism's probabilistic representation of the environment). Free energy forms an upper bound on the surprisal on sampling some data, given a generative model. And generative models (denoted by m) are probabilistic models of the dependencies between causes and effects from which samples can be generated (Friston, 2010, p. 2). Perception is a result of belief updating, under a generative model that comprises likelihoods and prior beliefs. The surprisal is formulated in terms of the negative log probability of an outcome, and entropy is the average surprisal formed on the basis of outcomes sampled from a probability distribution or density (Friston, 2010, pp. 1–2). The theory defines perceptions as "the results of Bayesian inversion of a causal model, and causal models are updated by the processing of sensory signals according to Bayes' rule" (Buckley, Kim, McGregor, & Seth, 2017, p. 56). The theory indicates that organisms have to keep their surprisal low, in order to maintain a nonequilibrium steady-state with their environment (i.e., generalised homoeostasis). To do so, the organisms have to generate reliable models of their environment and update their models and reduce their errors so that the models become reliable. Organisms, according to this approach, implement variational Bayesian inferences to determine the hidden state (or causes) of the world that underlie the (sensory) data (hence the relation between FEP and variational Bayesianism). Similarly, it is possible to use a Lyapunov function—a scalar function of a system's state that decreases with time—in order to infer the states and behaviours of a system that applies FEP (Friston, 2018; Friston & Stephan, 2007). Lyapunov functions can be used to model the stability of the system. Lyapunov function's formalism is in harmony with FEP which aims to account for the stability of the organism's dynamical relationship with its environment.

Organisms actively garner evidence for the correctness of their hypotheses (or models) about the world. Active inferences lie at the centre of FEP's account of the organism's representations and its relationship with the world. Motile organisms could garner evidence for their models of environment either by searching the environment and updating their internal models according to their findings, or alternatively by changing their environment so that it conforms to their internal models. In short, the organism's models of the environment could be constantly updated and optimised on the basis of the prediction error reduction mechanisms. These mechanisms aim to keep the organism in the state of equilibrium with the environment. This account lines up with the theoretical biology and the theory of evolution satisfactorily, because it indicates that organisms that are capable of reducing their prediction error (aka surprisal) can maximise their survival. It would even be possible to construe evolution itself as a process of Bayesian inference. In this setting, natural selection becomes a process of Bayesian model selection based upon the adaptive fitness that is scored by the surprisal accumulated by a phenotype. This treatment of evolution suggests that each phenotype becomes a hypothesis or prediction about the sort of creature that would be most apt for a particular eco-niche (Allen & Friston, 2016). Organisms that possess a good fit with their environment are therefore selected in virtue of having a low surprisal (or, mathematically, high Bayesian model evidence). This account places cognition within a hierarchical framework; where genetic and epigenetic priors are (literally) inherited from a similar process of Bayesian inference at an evolutionary scale.

FEP accounts not only for the organism's representation of its environment but also for the processing of the self-related stimuli and the organisms' interoceptive representations (Apps & Tsakiris, 2014; Gallagher & Daly, 2018; Limanowski & Blankenburg, 2013). It accounts for the generation of the phenomenal aspects of the self as well as the experience of body ownership, given that the organism's model of itself is based on its predictive representation of its own body (the location of the body, its morphology, etc.). Accordingly, it has been argued that self-recognition is a result of the brain's attempt at minimising its free energy by being in states where the environment is most predictable (Apps & Tsakiris, 2014, p. 86; Fotopoulou, 2012). Within this context, Friston (2018) explicates consciousness as a natural inferential process. However, he does not want to claim that all natural processes such as evolution or weather are conscious. Thus, his theory includes something quite close to a criterion of distinguishing models capable of accommodating conscious states and models incapable of it. The criterion is defined in terms of "temporal thickness" and relevant notions. In the next sub-section, I shall outline the FEP-based theory of consciousness.

5.4.2   The FEP-Based Theory of Consciousness

According to Friston (2018), consciousness is an inferential process. FEP and its Bayesian mechanisms provide a formal framework for modelling this process. I shall outline the FEP-based theory of consciousness (or rather self-consciousness) immediately.

Let us assume that, as Friston (2018) suggests, consciousness requires self-consciousness. Friston defines self-consciousness in the context of the organism's dynamical interrelationship with the environment. As I have already explained, Friston characterises the relationship between the organism and the environment under the rubric of FEP and active inferences. To engage in active inferences, the organism must be able to model the consequences of its action to itself (given that the consequences postdate the action). Models of organisms that can regiment the consequences of the actions are temporally thick, and "temporal thickness" underpins a sufficient account (or in my terminology, a demarcation criterion) of consciousness.

In developing his theory of consciousness, Friston provides Bayesian formulations of the connection between notions of "temporal thickness", "counterfactual depth", "purposefulness", "self-evidencing", and "agency". Organisms whose models are temporally thick are purposeful, self-conscious, and by the same token conscious. Friston's sufficient condition or demarcation criterion of consciousness consists in "temporal thickness" and "counterfactual depth" of the active inferences. He argues that only things with temporally thick models are capable of (self)consciousness. The behaviour of organisms with thick or deep models is epistemic, purposeful, and agentive. Models of such organisms are supposed to possess epistemic functions, and their inferences can underpin self-conscious states. However, it could be contended that notions of purposefulness, agency, and so on, and their formal counterparts (e.g., temporal thickness) do not provide objective grounds for discriminating between (what we recognise as) conscious and unconscious things. I shall spell out reasons for this scepticism immediately.

Some organisms can model the world to themselves on the basis of their beliefs about the future states of the world, not its present states (Friston, 2018, p. 579; Friston et al., 2015). Within this context, active inferences are defined as approximate Bayesian inferences and active sampling of data on the basis of the distribution of posterior beliefs over action, under the prior belief that action will minimise free energy in the future (Friston, 2018, p. 279). Active inferences aim to decrease uncertainty on the basis of the inferential mechanisms underpinning epistemic, information-seeking behaviour (Friston, 2018, p. 579; Friston et al., 2015). The Bayesian framework of this theory covers the notions of belief, uncertainty, agency, purposefulness, and temporal thickness. However, it could be contended that although these notions could be modelled with formal precision via variational Bayesian methods, they hardly provide insights into organisms' subjective states. For example, it can be contended that uncertainty and purposefulness can be defined with precision in the context of Bayesianism, but they do not refer to the organism's epistemic and mental states, nor do they model the organism's subjective experiences. Similarly, technical notions of uncertainty, purpose, and so on, as being explicated in the Bayesian framework, do not identify with the subjective quality (or feeling) of purposefulness, certainty, and so on. To use a well-known terminology of this field (coined by Nagel (1974)), from knowing Bayesian models of a given organism's (e.g., a bat's) perception, it does not transpire how it feels to be that particular organism. On the other hand, there is no way to know that things (e.g., a stone, a river) that allegedly do not have temporally thick models do not feel purposefulness and agency or lack subjective experiences. It follows that having temporally thick models is neither necessary nor sufficient for having the capacity for being self-conscious; and it does not provide an objective criterion for demarcating the vegan's mental states from processes realised by viruses, evolution, or economics. To be clear, it can be argued that, having temporally thick models is not an *objective* sufficient condition or demarcating criterion of consciousness, because it is possible to use temporally thick model instrumentally to describe the behavioural patterns of things that are not recognised as having the capacity for consciousness by Friston (e.g., evolution).

5.4.3 Realism and Antirealism About the FEP-Based Models of Consciousness

Please note that Friston takes a realist attitude towards the Bayesian thick models that represent the structure of consciousness. In the face of this realist attitude—which is in complete agreement with the structural *realist* trajectory of my own book—an antirealist may argue that despite their systematising capacity, models with temporal thickness and counterfactual depth cannot represent the objective structure of reality. Previously, an instrumentalist stance has been used to launch a criticism of the unificatory power of Bayesian approaches to psychology (FEP and predictive coding included) (Colombo & Hartmann, 2015; Colombo & Series, 2012). The same instrumentalist stance could underpin scepticism about the objectivity of notions of temporal thickness and counterfactual depth and their capacity in characterising the objective structure of consciousness. One way to flesh out this claim is to show that it is possible to ascribe purposefulness and agency to systems that Friston regards as incapable of consciousness (e.g., evolution). Below, I shall briefly spell out the problem.

Bayesianism does not provide a sufficient account—or a sufficient condition, a demarcating criterion, and so on—of cognition. Nor does it provide a viable criterion for demarcating conscious systems from unconscious ones. As Friston points out, FEP accommodates an ultimate deflationary account of everything (living things included), because every process that is measurable (i.e., has the states that are occupied frequently) should increase Bayesian model evidence in virtue of its very existence (Friston, 2018, pp. 579–580). So, if we consider consciousness in terms of basic capacity for subjective experience (as some authors did (Klein & Barron, 2016)) we must concede that, given Friston's definition, everything is basically conscious. I am quite ready to settle down for panpsychism, assuming that we could regard everything as conscious on the basis of a Bayesian criterion. However, Friston denies that all complex inferential systems such as evolution and virus are capable of being conscious. Friston's reasoning for this negative attitude may seem confusing. According to him:

> But is evolution really conscious? Probably not, for the following reason: previously, we have noted selection rests on processes embedded at multiple hierarchical levels—Darwinism within Darwinism all the way down to the selection of dendritic spines on single neurons in our brain (Edelman, 1993; Kiebel and Friston, 2011). At what point do these deeply entailed and hierarchical selection processes qualify as conscious? For example, a virus possesses all the biotic, self-organizing and implicitly inferential dynamics to qualify as a process of Bayesian inference; however, it does not have the same qualities as a vegan. (Friston, 2018, pp. 4–5)

What may seem confusing about Friston's statement is that he responds to the question (about the consciousness of evolution) with another question, which is supposed to indicate that viruses are not conscious either, because they do not have the same qualities as vegans. Despite this, Friston's statement betrays the main insight behind the enterprise for finding a sufficient account of consciousness, which results in spelling a demarcation criterion in terms of temporal thickness. The insight is based on the difference between things that resemble us (vegans) and those that do not resemble us (e.g., viruses, evolution). It holds that only things that resemble us are capable of being conscious. Let me elaborate.

If we concede specifying consciousness in terms of Bayesian inferences, we must grant that viruses and evolution are essentially conscious in the same way that vegans are, though perhaps there will be a difference in the degree of consciousness. However, our habituated intuitions revolt against this straightforward reply. To find a foothold for defending this insight, Friston takes a step further and suggests that self-consciousness is necessary for consciousness. Although Friston does not provide valid arguments to show how it is that the condition of self-consciousness is required by the condition of being a Bayesian inference (i.e., the primary condition of consciousness), I grant this point for the sake of argument. It would still be possible to argue that "temporal thickness" and similar notions that are invoked by Friston do not provide an objectively sufficient account or demarcating criterion of consciousness. Here, the antecedent of the antirealist argument about the FEP-based account of consciousness is that we can use temporally thick models to systematise the behaviour of things that are not normally recognised as conscious. Friston suggests that having models with temporal thickness (or counterfactual depth) is a property of self-conscious models, and given the soundness of the mentioned antecedent, it could be concluded that all things are conscious, because we can scribble down temporally thick models of them. Thus, it could be argued that Friston's theory supports panpsychism. On the other hand, someone who is keen to avoid panpsychism may conclude that, because we can make temporally thick models of all kinds of everything, the criterion of temporal thickness does not provide an objectively sufficient account or a criterion of consciousness at all. Let me elaborate.

According to Friston et al. (2015, p. 191 ff.), there is a distinction between organisms that could perform purposeful behaviours and those that cannot do so. The distinction relies on the point that only the former (but not the latter) kind of organisms can furnish beliefs about their future. Purposeful organisms' predictive posteriors include beliefs about future outcomes, counterfactual outcomes, and hidden states, whereas the current posterior of the other kinds of complex systems covers only hidden states (Friston et al., 2015, p. 192). The former kind of organisms' models is temporally thick. According to Friston et al. (2015, p. 191), this means that purposeful agents believe that they are minimising their surprisal, in addition to actually minimising it. However, it could be objected that all of these fine technical elaborations are conjured to serve our prejudice concerning the capacity of creatures like ourselves for pursuing their course of action consciously. To the extent that habituated intuitions are at issue, it does not make sense to say that evolution enjoys an intrinsic point of view, nor shall we say that evolution acts to increase epistemic value. This is because we *know* that evolution does not entertain counterfactual beliefs and uncertainty about the future. Neither does it make sense to understand a virus' environmental explorations in terms of its epistemic action. But these restrictions are not necessarily based on objective facts. For, it could be argued that the sophisticated Bayesian framework that is used by Friston is a mere theoretical instrument that serves the modeller's interests. That is to say, the Bayesian formalism itself does not provide an objective or factual demarcation criterion of consciousness. It is in principle possible to produce Bayesian models that characterise evolution as a purposeful process (i.e., as temporally thick models), although intuitively it does not make sense to say that evolution is capable of subjective experiences. Below, I show that it is indeed possible to model evolution as a thick model.

Arguably, evolution itself could be understood in terms of a Bayesian inference, where Bayesianism is supposed to be nature's way of "testing various hypotheses (i.e., models or phenotypes) and scoring them to select phenotypes that have the greatest evidence: adaptive fitness is just the evidence for the hypothesis that this phenotype can survive in this econiche" (Friston, 2018, p. 579). Does it make sense, then, to say that by increasing evidence, evolution minimises its uncertainty (as expected surprisal)? It would not be intuitively appealing to construe evolution and natural selection in epistemic terms. This is because we do not presume that evolution has a mind of its own. Nor does it make sense to ascribe things like metal states or subjective experiences to evolution and natural selection. According to Friston:

> we elude the problems of calling evolution conscious, because the process of natural selection minimizes surprisal (i.e., maximizes adaptive fitness) but not expected surprisal or uncertainty (i.e., adaptive fitness expected under alternative evolutionary operations or selection). The key difference between (self) consciousness and more universal processes then appears to be the locus of selection. In non-conscious processes this selection is realized in the here and now with selection among competing systems (e.g., phenotypes). Conversely, the sort of selection we have associated with (self) consciousness operates within the same system—a system that can simulate multiple futures, under different actions, and select the action that has the least surprising outcome. (Friston, 2018, p. 5)

Friston's reply indicates that natural selection takes place here and now (without any foresight into the future). This reply is in harmony with common sense. However, the intuitive appeal of this reply should not distort the philosophical point that the decision to model the dynamics of the relationship between the evolution and the world without assigning thick temporal models to evolution is a matter of the modeller's interests (meaning that it is not a fact of the matter). There is copious philosophical literature on the nature of models and model-making that can elucidate this point (Currie, 2016; Frigg, 2010; Godfrey-Smith, 2009). Even without delving into such philosophical discussions, it could be observed that the notion of temporal thickness is not (necessarily) indicative of the objective capacity for purposefulness and consciousness. This is because the modeller's intentions and presumptions bear on her decision to model things as thick or thin processes. Instead of rifling through the copious literature in the philosophy of science to flesh out my point, below I demonstrate it by showing that we could model evolution as a temporally thick system if we chose to do so.

Natural selection aims to select those traits that maximise survival and reproduction and transfer them from one generation of a population to the next. Traits that enhance fitness are inheritable, and various traits (expressed as behavioural, morphological, and physiological factors) cause various rates of survival. The process of natural selection is self-evidencing, in the sense that the survival of organisms is evidence for the truth or correctness of the natural selection's hypotheses about the traits that maximise survival. As I show immediately, we can model natural selection as a purposeful and agentive process. That is to say, it is possible to model natural selection as a process that forms expectations about the functional performance of traits that it chooses. It harbours expectations about the bearing of those traits on the fitness of organisms and the rate of their survival. Natural selection's choice of the types of traits that could enhance survival relies on its foresight into the predicted fitness of the organisms that inherit those traits. This means that the same formal tools that are suggested by Friston et al. (2015) for the purpose of modelling epistemic functions also model natural selection. Friston et al. suggest that "the negative expected free energy can also be expressed as the expected divergence between the posterior (predictive) distribution over hidden states with and without future observations, plus the expected utility (defined as the log of the prior probability of future states)"(Friston et al., 2015, p. 189). In the case of natural selection, too, we can show that minimising expected free energy about the fitness that a trait confers on types in a population reduces uncertainty about the survival of the type that inherits the trait (by maximising information gain). Statement of this construal of natural selection is based on a variational Bayesian framework (like what is introduced in Friston et al. (2015)), where the process not only relies on the explicit value of outcomes but also uses their epistemic value in reducing uncertainty about the relation between organisms and the environment. Clearly, we can characterise the epistemic value in terms of resolution of uncertainty about the functional performance of a given trait or its bearing on the survival rate in the descendant population. Minimising uncertainty would increase natural selection's confidence in the correctness of a hypothesis about the expected fitness or survival rate that a trait confers on a descendant population. If the trait decreased the fitness of the organism or minimised the survival of a population, natural selection would be surprised. Please note that I say we *can* characterise our construal of natural selection and evolution along those lines, without indicating that this is the right way to do so. Despite having a choice to do so, we do not actually endeavour to characterise or understand evolution in terms that I suggest here. That is to say, none of what I said in this paragraph makes sense because we (think we) know that natural selection does not have a mind of its own and does not rely on expected outcomes and counterfactual inferences to pursue its course on the basis of anything like epistemic insights. As Friston clarifies,

> Although it is possible that *"we could model evolution as a temporally thick system if we chose to do so"* this is not relevant. One would need to demonstrate that evolution was a particular kind of inference that could only be explained under a temporally thick model. In short, it is always possible to objectively *"model natural selection as an epistemic purposeful agent (with a temporally thick model)"*. However, this is not evidence that natural selection can be described as inference under a temporally thick model. (Friston, personal communication)

Thus, it is in principle possible to use Kullback-Leibler divergence to model fitness in terms of the divergence from the frequency of a type in the initial population to expectations[1] about the frequency of updated population after selection. To the extent that formalism is at issue, we could ascribe purposefulness, epistemic states, and expectations to evolution. However, although we may naturally assume that "selection is the process by which populations accumulate information about the environment" (Frank, 2012, p. 2392), we usually do not go so far as to model the change in the frequency of types through generation in terms of evolution's own uncertainty, confidence, and other epistemic states because there is no evidence that evolution uses thick models. Models can be forged instrumentally and instrumental formalism does not prevent us from ascribing purposefulness and agency to evolution. However, we do not choose to do so, because there is no positive evidence to indicate that evolution is capable of having foresight into the future. We use temporally thick models to describe the behaviour of organisms such as ourselves but not viruses, genes, or evolution. In short, we discriminate between viruses and evolution on the one hand and organisms such as ourselves on the other. We explain the behaviour of the latter group but not the former by attributing epistemic agenda and purposefulness to them. I do not intend to suggest that this discriminatory insight is completely groundless. My point is that the Bayesian formalism does not support the attempt at providing an objective criterion for demarcating creatures capable of having subjective experiences from complex systems that are deemed incapable of having such experiences.

In response to the mentioned criticism, Friston clarifies that, despite the possibility of using temporally thick Bayesian models to regiment evolution, this way of using the models is not what self-organisation, self-evidencing, and ensuing self-consciousness are based upon. While it could be granted that there are indeed certain aspects of natural selection that have a predictive and temporally extended aspect, the sort of anticipatory, instrumental processes that furnish evidence for temporally thick models do not indicate that evolution is self-conscious. This is because the FEP-based model is not based upon the possibility of forging thick models of processes by a modeller. It is based on the capacity of some

---

[1] Expectations can be characterised in terms of the prior beliefs about the frequency of the descendent population on the basis of the fitness of the ancestral population.

organisms for modelling their own futures. In this setting, epistemic value is an attribute of the counterfactual outcomes following the action of the organism; namely, actions that resolve uncertainty under beliefs about the consequences of an action for the organism itself. The model does not "*possess epistemic value*"; epistemic value is an attribute of a policy that is modelled. According to Friston,

> The question is not whether you or I can model evolution with a deep generative model—the question is: does evolution model itself with a temporally thick model? In other words, does the self-evidencing (Hohwy 2016) of evolution show any evidence that uses temporally thick models. Practically speaking, this would require some physical state of the evolutionary process to encode some aspect of policy selection and implicit (probabilistic) beliefs about the consequences of pursuing different policies. Put simply, there is no empirical evidence that evolution chooses between courses of action on the basis of a model of "what happen if I did that". (Schmidhuber 2010) (Friston, personal communication)

So indeed a modeller could use temporally deep models to model processes such as evolution. But from Friston's point of view this is beside the point about the capacity of the organism itself to model its future and use it as a guide to action. A modeller could use the notion of counterfactual depth to model processes that are even less likely to be self-conscious than evolution. For example, it is possible to use such models to describe the earth's travel around the sun, where the earth revisits the same state on a regular basis to garner evidence for its existence. Drifting away from the orbit would minimise the earth's survival, perhaps in the same way that getting out of water would minimise a fish's survival or in the way that submerging underwater would minimise a human's survival. However, we have no evidence to think that the earth's own revolution is purposeful and planned in the same way that an organism's attempt at maximising survival is. There may be a way for formalising this insight, by saying that the earth cannot choose its orbit—as a state that it occupies frequently—whereas an organism is capable of choosing its behavioural routines. A purposeful thing can be modelled as a "random dynamical attractor, with an attracting set of states that fills a large part of (state) space, yet has an intrinsically small volume (i.e., measure)" (Friston, 2018, p. 3). Of all possible states such a thing could occupy, there are only a small number of subspaces that it would characteristically be found in (ibid.). From an antirealist perspective, the important point is that the variational Bayesian formalism does not ipso facto prevent us from modelling the earth's revolution by invoking a random dynamical attractor, with an attracting set of states that fills a large part of (state) space (all the universe), yet has an intrinsically small volume (around the earth's orbit). Our reluctance to model the earth's revolution on such basis is not invoked by our formalism, but by our tendency about the earth's (or the universe's) purposefulness (or lack thereof). An antirealist may argue that this tendency is not based on objective facts about the nature of planetary movements or the epistemic states of planets (assuming that planets might possess such states). However, this does not seem to be the right way of understanding Friston's use of Bayesianism for modelling consciousness. According to Friston, to the extent that the example of planetary motion is at issue,

> to ascribe self-consciousness two heavenly bodies, one would have to find empirical evidence that a planet chose this orbit as opposed to that orbit. On this reading, the objective criteria for demarcating conscious and non-self-conscious processes (i.e., things) reduces to establishing evidence for or against an inference process that entails policy selection. From the perspective of machine learning, this is the difference between Bayesian inference and Bayesian decision theory. From the perspective of a neuroscientist, this would boil down to providing the neural correlates of consciousness as it pertains to intention and choice about "what I am going to do next". In this view, there is an objective criterion for demarcation, based upon the simplest explanation for the behaviour of any observable system. (Friston, personal communication)

In this sense, Friston's use of Bayesian models is in agreement with our commonplace discriminatory intuitions, which dissuade us from ascribing consciousness to things as dissimilar to ourselves as evolution or planetary movements. Based on his clarification, his approach may successfully sidestep the antirealist criticism concerning the plausibility of the FEP-based theory of consciousness. On the other hand, it could be still remarked that the capacity to attribute consciousness to some entities but not to others is itself a subjective capacity possessed by some organisms (such as ourselves) that have self-consciousness. Below, I will unpack this last remark.

Please note that defending the realist credentials of the FEP-based theory does not mandate corroborating the theory against the charge of being accompanied by a discriminatory insight. As Friston acknowledged, from the perspective of Bayesian mechanisms, "if I exist in a world that is populated with other creatures like me, then I will come to learn this fundamental state of affairs [...]. [...] most of my generative model is concerned with modelling you, under the assumption that you are a "creature like me"" (Friston, 2018, p. 7). And to the extent that my generative model's insight into your consciousness is based on the measure of your resemblance to myself, the model relies on the discriminatory bias. The capacity for having subjective experiences is a feature of my (i.e., the modeller's) generative models that I project into other systems (that may resemble me). I shall explain this last point immediately. Given the bias of our generative models (which ascribe the capacity for being conscious on the basis of similarity of other things to us), we are not supposed to model evolution in terms of epistemic behaviour, mainly because evolution is not a creature like ourselves. We can analyse the notions of purposefulness and agency in the same way, by saying that ascribing agency, purposefulness, and goal-directedness to some systems but not others is a matter of the modeller's explanatory interests and choices. And the modeller's decisions might be systematically biased by the modeller's partiality to the resemblance of some target systems (but not others) to the modeller's own generative models. We (as modellers) project purposefulness and agency to complex systems that resemble us. There is some discriminatory intuition that makes us consider systems that are like ourselves as capable of sporting conscious states. However, there is no reason to think that this discriminatory attitude is not based on facts of the matter or objective constitution of the real world. The discriminatory categorisation of the world into things that are either capable or incapable of having conscious states may be an inexorable way of understanding the environment for organisms like us. Our generative models work in the way that they do, and philosophical enlightenment as regards their discriminatory strategy cannot change the way that we use them to perceive the world. However, acknowledging this discriminatory base does not need to undermine the realist core of the FEP-based model of (self)consciousness. As Friston argues,

> I think it is perfectly sensible to understand a discriminatory (and possibly unwarranted) tendency to ascribe self-consciousness to "creatures like me". It is interesting to consider whether this is a necessary state of affairs. In other words, if self-consciousness and subjective "feelings" are themselves parts of our generative models then, a priori, it is plausible that only models with this particular capacity can infer that there are "creatures that have feelings". This simple point here is that qualitative experiences and "feelings" may themselves be a big part of a portfolio of hypotheses or explanations that provide the simplest explanation for our sensed body and cogitation. (Friston, personal communication)

In the next section, I will elaborate on the structural relationship between various theories of consciousness that have been discussed thus far in this chapter.

5.5 On the Structural Relationship Between the Three Theories of Consciousness

In this section, I account for the connection between the FEP-based account of consciousness on the one hand and the resting-state-based theory and IIT on the other. Note that, as I have remarked in Sect. 5.1, the philosophical conception of consciousness is underdetermined by the diversity of theoretically and empirically adequate theories that claim to provide plausible accounts of the underpinning mechanisms of cognition. In the previous sections of this chapter, I outlined three of the most important theories of consciousness. These theories have been presented in different contexts. As I have already remarked, IIT is mainly a mathematical theory which characterises the properties of conscious experiences. The resting state account is mainly concerned with the neurological bases of consciousness and FEP provides a Bayesian account that has its stronghold in theoretical biology. Which one of them is the correct theory of consciousness? A structural realist theory of consciousness does not need to choose one of these theories as the exclusively correct theory of consciousness. These theories describe various parts of the underpinning structure of consciousness. According to my realist version of the structuralist theories of consciousness, the fact that there are structural relations between the mentioned theories indicates that there are overlaps between various parts of the underpinning patterns of consciousness. That is to say, the existence of commonalities between various theories indicates that the structures that are described by the three mentioned theories are indeed pieces of the same unified whole or underpinning structure which exhibits various phenomenal properties of the self. In the remainder of this section, firstly, I describe the relationship between the FEP-based account of consciousness and the resting-state-based theory. Then I shall explicate the connection between the FEP-based theory and IIT. Unveiling the existing structural relations could contribute to defending a unified structuralist theory of consciousness.

Let us see how the FEP-based theory connects with the resting-state-based theory. At first glance, there are divergences between these two theories. FEP provides a first principle account of systems that tend to minimise their free energy in order to maximise their survival, whereas the difference-based coding concerns basic kind of neural activity that makes the brain's processing of interoceptive and exteroceptive information possible. These theories are describing two different ways of neural coding. The FEP-based account of consciousness is based on the brain's capacity to minimise its free energy by reducing its prediction error, whereas the resting-state-based theory presumes that consciousness is based on the resting state activity realised by the brain's capacity to encode the differences between discrete spatial and temporal properties. Despite the discrepancies, it is possible to account for compatibility of the FEP-based and resting-state-activity-based accounts of consciousness. This claim could be substantiated by showing that these two kinds of coding (i.e., predictive coding and difference-based coding) are compatible. Here, the general insight is that it is possible to assume that predictive coding presupposes difference-based coding and is of a special kind. Georg Northoff has substantiated this point duly in his account of the resting state theory and difference-based coding (Northoff, 2014a, part III).

As Northoff (ibid.) has argued, predictive coding is based on minimising the prediction error, which is basically a discrepancy or difference between the set of predicted inputs and the set of actual inputs. This means that the neural activity that realises predictive coding is based on the brain's capacity to mark a difference between expected and actual inputs. Predictive coding is a result of the brain's capacity for comparing these two sorts of inputs. In this sense, predictive coding is difference-based, because it is based on encoding and comparing the difference. Thus difference-based coding is presupposed by predictive coding (Northoff, 2014b, pp. 63–64). It is also worth mentioning that the technical explication of FEP, for instance, its use of Kullback-Leibler divergence as an information-theoretic measure that characterises the divergence between the probability distributions in the expected and actual sets of inputs (see Sect. 5.4 in this chapter) does support Northoff's claim about the reliance of predictive coding on difference-based coding—for the explication of FEP and predictive coding presumes that the organism (or its cognitive system) can discern the divergences and minimise them. Thus, the basic formulation of FEP is consistent with Northoff's general assumption, according to which the difference-based activity or encoding differences between discrete points in space and time is the brain's fundamental model of information processing. Northoff's theory reveals the fundamental mode of the brain's information processing. Given this assumption, I follow Northoff to submit that the FEP-based account of self-consciousness is consistent with the difference-based coding and presupposes it.

Perhaps it is worth mentioning that, in order to substantiate his point about the relationship between predictive coding and difference-based coding, Northoff draws on neuroimaging studies of sensorimotor mechanisms that underpin predictive coding. These studies arguably support the claim that predictive coding relies on difference-based coding rather than stimulus-based coding (Northoff, 2014a, p. 145 ff.). Also, note that being underpinned by difference-based coding vindicates the structuralist credentials of the FEP-based theory of consciousness (if there are any doubts about its structuralist credentials at all)—for the FEP-based theory is based on the organism's capacity for encoding the difference between discrete points in space and time, that is, the web of relations between predicted and actual inputs, rather than the relata or intrinsic properties of the discrete points themselves. This paves the way to a structuralist understanding of the FEP-based theory, which gives the relations and structures priority over relata and intrinsic properties of individual entities.

The FEP-based theory can also be related to IIT. It is true that there may be some discrepancies between FEP and IIT; FEP provides variational Bayesian models of organisms that could embed deep models of their future, whereas IIT provides an axiomatic formulation of features of conscious experience. However, as Friston, Weise, and Hobson (forthcoming) have recently argued, it is possible to build upon the information-theoretic core of the FEP-based theory and IIT to ground the compatibility of these two theories. This is because arguably minimising variational free energy (in the FEP-based account) corresponds to maximising "phi" (in IIT) and vice versa (for technical details, see Friston et al., forthcoming). This indicates that FEP can provide Bayesian generative models of types of conscious experiences that are supposed to be characterised in IIT's axiomatic system (the axioms are mentioned in Sect. 5.2). If we grant Friston et al.'s (ibid.) account of the interconnection between the FEP-based theory and IIT, we can conclude that there are structural (information-theoretic) interconnections between them, and this adds up to the conclusion that the FEP-based theory and IIT can be subsumed by a unified structural account of consciousness. This provides an important purchase for defending the plausibility of the structural realist theory of consciousness. As I have argued in Chaps. 2, 3, and 4, structural realist theories of various stripes seek to substantiate their ontological plausibility by showing that common structures underpin apparent theoretical diversities that wreak havoc with the metaphysical plausibility of some theories (this has been formulated in this book and elsewhere as the problem of metaphysical underdetermination). Now, showing that the diverse theoretical accounts of consciousness that are described in this chapter— that is, IIT, the resting-state-based account, and the FEP-based account— are structurally related together, indicates that there is structural unity underpinning the theoretical diversity of neurophilosophical account of consciousness. The structural realist theory of consciousness is committed to the unifying structure that underpins diverse theories and integrates them. The fact that it is possible to draw formal relations between various theories consolidates the philosophical plausibility of a unifying account of the structural realist theory of consciousness. Before ending this chapter, I also present a structural realist account of another (though somewhat relevant) phenomenal aspect of the self.

5.6 Structuralism About Intentionality

It is possible to assume that at least some subjective experiences are directed towards some objects. "Intentionality" can be specified in terms of the capacity of mental states to be about something, to represent the world's states of affairs, and to bestow upon experiences their phenomenal character. Some philosophers assume that there is a close connection between consciousness and intentionality. Franz Brentano, for example, presumed that intentionality is the mark of all mental phenomena, consciousness included. According to Brentano ([1874] 1995, pp. 88–89),

> Every mental phenomenon is characterized by what the Scholastics of the Middle Ages called the intentional (or mental) inexistence of an object, and what we might call, though not wholly unambiguously, reference to a content, direction towards an object (which is not to be understood here as meaning a thing), or immanent objectivity. Every mental phenomenon includes something as object within itself, although they do not all do so in the same way. In presentation something is presented, in judgement something is affirmed or denied, in love loved, in hate hated, in desire desired and so on. This intentional in-existence is characteristic exclusively of mental phenomena. No physical phenomenon exhibits anything like it. We could, therefore, define mental phenomena by saying that they are those phenomena which contain an object intentionally within themselves.

As this quote indicates, Brentano characterises all mental phenomena in terms of intentionality. Tim Crane has restored and advocated this account of intentionality (Crane, 1998, 2001). There are yet others, such as John Searle (1983), who assume that there is an intimate relationship between consciousness and intentionality, without characterising the former in terms of the latter—instead, Searle argues that the consciousness is the mark of the mental. While this section is concerned with intentionality, I do not defend the thesis that consciousness or subjective experiences are necessarily associated with intentionality or have an intentional aspect. So, a fine line has been drawn between intentionality and phenomenality before (Block, 1995; Searle, 1983), and it is not my goal to demonstrate that the two are the same or even strongly connected. My endeavour in this section needs to presuppose only a lax relationship between consciousness and intentionality. By speaking of "*lax* relationship", I submit that all information processing (*at least to some extent*) involves phenomenal experience. I do not engage in defending this view here, the view seems to follow from my attempt at presenting an information-theoretic account of the structure of the self and its basic properties. I understand that this submission can lead to the charge of pancomputationalism-cum-panpsychism, that is, the view that all systems that are capable of computation or information processing can have subjective experiences. At the risk of sounding maverick, I think I will plead guilty to the charge of advocating a *limited* form of panpsychism. I think this is in line with IIT that I have outlined in the earlier sections of this chapter. Also, this submission is in harmony with what I said in the previous chapter (Sects. 4.6 and 4.7) about the extension of the informational structure of the (embodied) self to its environment, for even the environmental informational structures should be capable of functioning as vehicles of phenomenal experience if the self and its environment are supposed to form a coupled dynamical system. However, I recognise only a limited form of pancomputationalism. Again, at the risk of sounding tautological, I accept that some forms of consciousness and intentionality (that are familiar to us) are only available to organisms such as ourselves (that are capable of self-consciousness), that is, organisms that are capable of constructing temporally thick, counterfactual models of themselves and their decision in their environment. (To see how intentionality could be explained in terms of the organism-environment dynamical relation, or trade-off of energy, see Khachouf, Poletti, & Pagnoni, 2013; Tschacher & Haken, 2007.)

Be that as it may, the general insight of this section is that, assuming that intentionality is an important mental capacity which is somehow (even if loosely) related to consciousness, complementing a structuralist theory of the self by a structuralist theory of intentionality would generally contribute to consolidating SRS. In this section, I build upon a previous enterprise (Beni, 2018b) to show how it is that a structural realist theory of consciousness addresses an enduring problem in the philosophical accounts of intentionality. The enduring problem is the problem of intentional inexistence.

As my previous reference to Brentano indicates, he mentioned the notion of "intentional inexistence". Let us proceed by considering the question of how to understand the (in)existence of objects of intentionality. At first glance, one might be tempted to identify intentional objects with ordinary objects. The object of my thought when I am thinking about a boat should be somewhat similar to the boat itself and even identified with it. The problem for this understanding of the objects of thought is that, unlike the ordinary objects, the intentional objects are not facts of the matter and might not exist at all. All else being equal— when I have normal experiences—the cat that climbs the tree in front of me certainly exists, whereas the winged horse that I am thinking about does not exist. So, objects of thought or intentional acts may be non-existent. A new vicious problem raises its head; to think about something consists in being in a certain relationship with it, and we cannot be in relationship with something that does not exist at all (Kriegel, 2007). Thus, the question of intentional inexistence emerges.

To account for the distinction between ordinary objects and intentional ones, Brentano developed the notion of intentional inexistence. There are various ways of understanding this notion. One may assume that the object of the thought is internal to the act of thinking, and "inexistence" is supposed to highlight the interdependence between the act of intentionality (or the relation of intentionality) and its object (or the relata). According to this construal, the object of thought is included in the act of intentionality like a relatum in the intentional relationship, and *in*existence is supposed to draw attention to this inclusion or interdependence. But not all objects of thought actually exist, and there may be non-existent objects of thought. Here, the question is, how can something that does not exist be included in a relationship? As Brentano submitted,

> The phenomena of light, sound, heat, spatial location and locomotion which [the scientist] studies are not things which really and truly exist. They are signs of something real, which, through its causal activity, produces presentations of them. They are not, however, an adequate representation of this reality, and they give us knowledge of it only in a very incomplete sense. (Brentano, 1995, p. 14)

Intentional objects do not exist, and thus the question of the relationship with something that does not exist raises its head. According to Tim Crane's statement of the problem, "how can a non-existent entity like Pegasus [a mythological winged horse] be the object of an act of thought, since it cannot be something which stands in relation to the subject of a mental act, because anything which stands in a relation to anything else must exist" (Crane, 2014, p. 29). The problem could be especially intriguing for a realist, given that the realist stance presumes that things to which we make ontological commitments do really exist. Because there could not be a real relationship with a thing that does not exist, Crane rejects the relationalist reading of Brentano's thesis. Crane's (2014) denial is based on the good old view about the impossibility of relatum-less relations. This problem has been also voiced against SR in the philosophy of science by some renowned critics of the theory (e.g., Psillos, 2006). Crane's argument is based on the same orthodox metaphysical intuition that dismisses the assumption of the ontological priority of relations over relata. In defence of this orthodox insight, Crane builds upon the antecedent that "the existence of a perceptual experience does not entail the existence of its object", to conclude that "experiences are not relations to the objects of experience" (Crane, 2014, p. 206). The problem with this reply is that the argument is not valid, of course, unless one presumes the impossibility of relata-less relations in the first place. If one presumes the impossibility of the relata-less relations, then the argument would become fallacious because its argument would beg the question of the impossibility of such relations (Beni, 2018b). My structural realist theory of intentionality, which will be unfolded in the next sub-section, does not comply with the orthodox view on the priority of entities over relations.

According to Crane, what makes assertions about non-existent objects or objects of thought true consists of facts about representation—for example, psychological states or episodes—and not the objects of thought as things that may not exist at all (Crane, 2013, p. 168). Crane's reply is based on the dichotomy between pleonastic and non-pleonastic properties. Pleonastic properties are relations that feature in the valid inferences, albeit without being ontologically constitutive. On the other hand, non-pleonastic properties are substantial entities which can feature in the ontology (Crane, 2013, p. 65). Crane builds upon this distinction to argue that pleonastic properties constitute non-existent objects which could be identified mainly on account of their place in the representational web. There is room for reasonable doubt about the plausibility of Crane's account. For example, one may doubt that it is possible to exhaust non-existence in terms of pleonastic properties and relations (see Jacob, 2014). Moreover, since non-existent objects have to be explicated in terms of intentional acts and relations which include the non-existent objects of thought, it can be claimed the account of intentionality and the theory of non-existent objects go hand in hand. Crane himself somewhat recognises this last point, by asserting that his account is somewhat circular (Crane, 2013, p. 168). This means that the questions of the nature of the intentional relations and the intentional content (as components of the theory) need to be addressed first (also see Beni, 2018b). Only then, and after addressing the mentioned point, we can proceed to understand the interdependence between intentionality and its objects. I shall unpack this remark in the next section. Below I shall spell out my own structural realist account of the intentional objects and their mode of existence.

One way to develop a structural realist theory is to begin from orthodox scientific realism. We can begin by submitting that full-fledged scientific realism is a reasonable stance to hold. The full-fledged version of scientific realism indicates that the world (i.e., what there is) exists precisely in the same way that is described by our theories. However, this view is unnecessarily presumptuous. We know that the world is not precisely in harmony with our theoretical descriptions because our theories can change and progress through their historical course, whereas the structure of the world is presumably fixed. That is to say, the change from the phlogiston theory to the modern theory of oxidation is not associated with changes in the nature of reality. So, after embracing the overall viability of scientific realism, we may want to give up with the presumptuous component of scientific realism and make concession on a more modest version of realism. In this fashion, we may take a downward path to realism and be realist only about the structure that underlies theoretical changes (Beni, 2018c). The result is structural realism (SR). I apply the same strategy to construct an intentional structural realist account of intentionality. To do so, I begin from a full-fledged realism about intentionality. This full-fledged version of realism includes both relations and objects of intentional acts. However, it is not plausible to profess realism about the intentional objects, owing to the problem of intentional inexistence, and given the fact that such objects do not exist. It follows that we might profess realism about intentional relations without going so far as to acknowledge the existence of the intentional objects, as things whose existence does not depend on the intentional relations. I think this view is metaphysically appealing because we can grant that an intentional object exists in virtue of being included within an intentional relation. This can provide the basis for a structural realist theory of intentionality that makes ontological commitments to the underlying structure of intentionality, without also being committed to the ontological priority of individual intentional objects.

To steer clear from being entrapped in the puzzle of the ontological status of non-existent objects, we must avoid making ontological commitments to intentional objects (in the sense of orthodox individual objects). Under the circumstances, it would be best to dispense with them and be a realist only with regard to the underlying structure of intentionality. Please note that I am developing the structural realist theory of intentionality along the lines of a non-eliminativist version of SR (NSR). I mentioned NSR in Chap. 2 when I was outlining various versions of SR. NSR is a version of SR that, despite giving priority to basic ontological structures, retains individual objects but evinces that they can be identified in virtue of their place in the structures. NSR allows for the existence of weakly discernible individual objects (or relata), which feature in the structures but assume that these objects are weakly identifiable on account of their place in the structures (Ladyman, 2007). By drawing on examples for relativity physics and quantum mechanics, some structural realists such as Esfeld and Lam have argued that it is not possible to draw a line within a scientific theory between the formal description of the structure and the description of intrinsic properties of the relata that feature in the structure (Esfeld & Lam, 2008, p. 29). It was in this spirit that they observed that the "[r]elations require relata, that is, objects that stand in the relations" (Esfeld & Lam, 2008, p. 29). Despite acknowledging the role of relations, NSR does not make full-fledged ontological commitments to the relata. As Esfeld and Lam have remarked, "*It is not the case* that these objects necessarily have intrinsic properties over and above the relations that they bear to one another" (ibid., emphasis original). Similarly, my structural realist view of intentionality indicates that there is no clear distinction between the intentional relations and the intentional objects in the theory of intentionality. The structural realist theory of intentionality retains a dwindled notion of objects of thought as what could be identified on account of their role as relata of the relations in the structure of the thought. While intentional objects exist, their existence is not independent of or prior to their role in the intentional structures. It is true that unlike Crane, I am making ontological commitments to the intentional relations. However, the problem of relata-less relations—as a serious obstacle to the relationalist account—does not need to arise here. This is because NSR leaves enough room for a thin notion of relata, despite draining them from their intrinsic properties, for intentional relations are not totally without relata, and the object of the intentional act does not need to transcend the act. Intentional objects are not eliminated. Rather, the point is that relata (i.e., the intentional objects) are simply devoid of any intrinsic properties beyond what could be ascribed to them on account of their role in the structure.

As I explained in Chap. 2, SR includes epistemological and ontological commitments to the structure of the theories, without going so far as to also commit to the existence of the individual objects with intrinsic properties. According to SR, it is reasonable to assume that our best scientific theories are informative about the structures in the enduring objective domain, without assuming that the individual objects that feature in the structures are knowable/exist independently of their underpinning structures. This proposal lines up with the original statement of Brentano's relationalist approach. According to Brentano,

The phenomena of light, sound, heat, spatial location and locomotion which he studies are not things which really and truly exist. They are signs of something real, which, through its causal activity, produces presentations of them. They are not, however, an adequate representation of this reality, and they give us knowledge of it only in a very incomplete sense. We can say that there exists something which, under certain conditions, causes this or that sensation. *We can probably also prove that there must be relations among these realities similar to those which are manifested by spatial phenomena shapes and sizes. But this is as far as we can go. We have no experience of that which truly exists, in and of itself, and that which we do experience is not true. The truth of physical phenomena is, as they say, only a relative truth.* (Brentano, 1995, p. 14, emphasis added)

As the quotation indicates, Brentano's view meshes nicely with SR, according to which what we could know about the world is its structure. To the extent that the structure of the perceptions is stimulated by the causal structure of the world, it could be informative about the spatio-temporal structure of the universe. However, as Brentano acknowledged, we have no access to the entities that inhabit the region of reality in itself. Intentional objects are only weakly discernible, and they can be identified regardless of their intrinsic nature, and only in virtue of their place within the web of intentional structures. The relata can be identified on account of their role in the structure intentionality, and are inseparable from the relations that characterise them. Thus the intentional objects are immanent to their structure (i.e., they do not completely transcend it) and are inseparable from it.

Let us recap. The structural realist account of intentionality satisfies some of the other intuitions that gave rise to Brentano's theory and Crane's construal of it. For example, the structural realist theory preserves the original interdependence of intentional relation and non-existent objects of thought completely. This theory can account for the ontic dependence between the objects of thought and intentionality in terms of the interdependence between relations and relata, where relata could be identified in virtue of their place in the structures. Unlike Crane's (2013) theory, the structural realist theory does not need to contrive a reductionist strategy for accounting for truths about non-existent objects in terms of truth about the existent phenomena, for instance, psychological states or processes. By the same token, it does not need to exhaust non-existence in terms of pleonastic properties and relations. The object of thought could be described as a relatum that features in the structure of the thought. The relatum does not possess any intrinsic properties beyond what is ascribed to it on account of its role in the structure of the relations that exhibit it. Thus, the structural realist theory of intentionality provides a viable reply to the question of the identity of inexistent objects.

5.7 Concluding Remarks

In this chapter, I introduced three structuralist accounts of consciousness. These were IIT, the resting-state-based account, and the FEP-based account of consciousness. As I remarked, all of these theories offer a structural characterisation of consciousness. They aim to specify the underpinning informational or embodied structures that exhibit phenomenal aspects of the self, without elaborating on the intrinsic nature of the phenomenal aspects. While the fact the theories can be construed along the structuralist lines provides some support for SRS, their diversity might distract from defending a structural realist theory of consciousness. For, the diversity could be understood as a token of lack of a general unifying framework that can subsume various theoretical and ontological implications of theories. Consequently, the problem of metaphysical underdetermination (mentioned in Chaps. 2, 3, and 4) might arise. To prevent this eventuality, I showed that it is indeed possible to reach a unifying account of the structural realist theory of consciousness, by elaborating on structural relations between three theories that are mentioned in this chapter. Thus, we may overcome the problem of metaphysical underdetermination of consciousness by the diversity of the structuralist theories of consciousness. The structural realist theory of consciousness is committed to the underpinning structure of the consciousness or the structure that exhibits the commonality between diverse theoretical descriptions of consciousness. I also offered a structural realist account of another significant phenomenal aspect of the self, that is, intentionality. I showed that it is possible to offer a unified structural account of important features of the self. And this provides further ground for being optimistic about prospects of SRS.

References

Adams, R. A., Shipp, S., & Friston, K. J. (2013). Predictions Not Commands: Active Inference in the Motor System. *Brain Structure and Function, 218*(3), 611–643. https://doi.org/10.1007/s00429-012-0475-5

Allen, M., & Friston, K. J. (2016). From Cognitivism to Autopoiesis: Towards a Computational Framework for the Embodied Mind. *Synthese*, 1–24. https://doi.org/10.1007/s11229-016-1288-5

Apps, M. A. J., & Tsakiris, M. (2014). The Free-Energy Self: A Predictive Coding Account of Self-Recognition. *Neuroscience & Biobehavioral Reviews, 41*, 85–97. https://doi.org/10.1016/J.NEUBIOREV.2013.01.029

Armstrong, D. M. (1968). *A Materialist Theory of the Mind*. London: Routledge & Kegan Paul.

Barrett, A. B. (2014). An Integration of Integrated Information Theory with Fundamental Physics. *Frontiers in Psychology, 5*, 63. https://doi.org/10.3389/fpsyg.2014.00063

Beni, M. D. (2018a). A Structuralist Defence of the Integrated Information Theory of Consciousness. *Journal of Consciousness Studies, 25*(9–10), 75–98. Retrieved from http://www.ingentaconnect.com/contentone/imp/jcs/2018/00000025/f0020009/art00003

Beni, M. D. (2018b). Much Ado About Nothing: Toward a Structural Realist Theory of Intentionality. *Axiomathes*. https://doi.org/10.1007/s10516-018-9372-8

Beni, M. D. (2018c). The Downward Path to Epistemic Informational Structural Realism. *Acta Analytica, 33*(2), 181–197. https://doi.org/10.1007/s12136-017-0333-4

Block, N. (1995). On a Confusion About a Function of Consciousness. *Behavioral and Brain Sciences, 18*(2), 227–247. https://doi.org/10.1017/S0140525X00038188

Brentano, F. (1995). *Psychology from an Empirical Standpoint* (L. L. McAlister, Ed.). London: Routledge.

Buckley, C. L., Kim, C. S., McGregor, S., & Seth, A. K. (2017). The Free Energy Principle for Action and Perception: A Mathematical Review. *Journal of*

Mathematical Psychology, *81,* 55–79. https://doi.org/10.1016/J.JMP. 2017.09.004

Colombo, M., & Hartmann, S. (2015). Bayesian Cognitive Science, Unification, and Explanation. *The British Journal for the Philosophy of Science, 68*(2), axv036. https://doi.org/10.1093/bjps/axv036

Colombo, M., & Series, P. (2012). Bayes in the Brain—On Bayesian Modelling in Neuroscience. *The British Journal for the Philosophy of Science, 63*(3), 697–723. https://doi.org/10.1093/bjps/axr043

Crane, T. (1998). Intentionality as the Mark of the Mental. In A. O'Hear (Ed.), *Contemporary Issues in the Philosophy of Mind* (Vol. 43, pp. 229–251). Cambridge: Cambridge University Press. https://doi.org/10.1017/S1358246100004380

Crane, T. (2001). Intentional Objects. *Ratio, 14*(4), 336–349. https://doi.org/10.1111/1467-9329.00168

Crane, T. (2013). *The Objects of Thought.* Oxford: Oxford University Press. https://doi.org/10.1093/acprof:oso/9780199682744.001.0001

Crane, T. (2014). *Aspects of Psychologism.* Cambridge, MA: Harvard University Press.

Cummins, R. C. (1996). *Representations, Targets, and Attitudes.* Cambridge, MA: MIT Press.

Currie, G. (2016). Models as Fictions, Fictions as Models. *The Monist, 99*(3), 296–310.

Dehaene, S., Kerszberg, M., & Changeux, J.-P. (2006). A Neuronal Model of a Global Workspace in Effortful Cognitive Tasks. *Annals of the New York Academy of Sciences, 929*(1), 152–165. https://doi.org/10.1111/j.1749-6632.2001.tb05714.x

Esfeld, M., & Lam, V. (2008). Moderate Structural Realism About Space-Time. *Synthese, 160*(1), 27–46. https://doi.org/10.1007/s11229-006-9076-2

Fotopoulou, A. (2012). Towards a Psychodynamic Neuroscience. In *From the Couch to the Lab* (pp. 25–46). Oxford University Press. https://doi.org/10.1093/med/9780199600526.003.0003

Frank, S. A. (2012). Natural Selection. V. How to Read the Fundamental Equations of Evolutionary Change in Terms of Information Theory. *Journal of Evolutionary Biology, 25*(12), 2377–2396. https://doi.org/10.1111/jeb.12010

Frigg, R. (2010). Models and Fiction. *Synthese, 172*(2), 251–268. https://doi.org/10.1007/s11229-009-9505-0

Friston, K. J. (2010). The Free-Energy Principle: A Unified Brain Theory? *Nature Reviews. Neuroscience, 11*(2), 127–138. https://doi.org/10.1038/nrn2787

Friston, K. J. (2018). Am I Self-Conscious? *Frontiers in Psychology, 9*, 579. https://doi.org/10.3389/FPSYG.2018.00579

Friston, K. J., & Kiebel, S. (2009). Predictive Coding Under the Free-Energy Principle. *Philosophical Transactions of the Royal Society of London. Series B, Biological Sciences, 364*(1521), 1211–1221. https://doi.org/10.1098/rstb.2008.0300

Friston, K. J., Rigoli, F., Ognibene, D., Mathys, C., Fitzgerald, T., & Pezzulo, G. (2015). Active Inference and Epistemic Value. *Cognitive Neuroscience, 6*(4), 187–214. https://doi.org/10.1080/17588928.2015.1020053

Friston, K. J., Samothrakis, S., & Montague, R. (2012). Active Inference and Agency: Optimal Control Without Cost Functions. *Biological Cybernetics, 106*(8–9), 523–541. https://doi.org/10.1007/s00422-012-0512-8

Friston, K. J., & Stephan, K. E. (2007). Free-Energy and the Brain. *Synthese, 159*(3), 417–458. https://doi.org/10.1007/s11229-007-9237-y

Friston Karl J., Wiese Wanja, and Hobson J. Allan Markovian Monism: From Cartesian Duality to Riemannian Manifolds.

Gallagher, S., & Daly, A. (2018). Dynamical Relations in the Self-Pattern. *Frontiers in Psychology, 9*, 664. https://doi.org/10.3389/fpsyg.2018.00664

Godfrey-Smith, P. (2009). Models and Fictions in Science. *Philosophical Studies, 143*(1), 101–116. https://doi.org/10.1007/s11098-008-9313-2

Greicius, M. D., Krasnow, B., Reiss, A. L., & Menon, V. (2003). Functional Connectivity in the Resting Brain: A Network Analysis of the Default Mode Hypothesis. *Proceedings of the National Academy of Sciences, 100*(1), 253–258. https://doi.org/10.1073/pnas.0135058100

Jacob, P. (2014). *Review of Tim Crane's the Objects of Thought*. Retrieved February 17, 2018, from https://ndpr.nd.edu/news/the-objects-of-thought/

Joshi, N. J., Tononi, G., Koch, C., Bonner, J., McShea, D., Adami, C., … Dijkstra, E. (2013). The Minimal Complexity of Adapting Agents Increases with Fitness. *PLoS Computational Biology, 9*(7), e1003111. https://doi.org/10.1371/journal.pcbi.1003111

Khachouf, O. T., Poletti, S., & Pagnoni, G. (2013). The Embodied Transcendental: A Kantian Perspective on Neurophenomenology. *Frontiers in Human Neuroscience, 7*(611). https://doi.org/10.3389/fnhum.2013.00611

Klein, C., & Barron, A. (2016). Insects Have the Capacity for Subjective Experience. *Animal Sentience: An Interdisciplinary Journal on Animal Feeling, 1*(9). Retrieved from https://animalstudiesrepository.org/animsent/vol1/iss9/1

Koch, C. (2012). *Consciousness: Confessions of a Romantic Reductionist*. Cambridge, MA: MIT Press.

Koch, C., & Tononi, G. (2013, March 7). Can a Photodiode Be Conscious? *The New York Review of Books*. Retrieved from http://www.nybooks.com/articles/2013/03/07/can-photodiode-be-conscious/

Kriegel, U. (2007). Intentional Inexistence and Phenomenal Intentionality. *Philosophical Perspectives, 21*(1), 307–340. https://doi.org/10.1111/j.1520-8583.2007.00129.x

Ladyman, J. (2007). On the Identity and Diversity of Objects in a Structure. *Proceedings of the Aristotelian Society, Supplementary Volumes*. Oxford University Press The Aristotelian Society. https://doi.org/10.2307/20619100

Lamme, V. A. F. (2006). Towards a True Neural Stance on Consciousness. *Trends in Cognitive Sciences, 10*(11), 494–501. https://doi.org/10.1016/j.tics.2006.09.001

Limanowski, J., & Blankenburg, F. (2013). Minimal Self-Models and the Free Energy Principle. *Frontiers in Human Neuroscience, 7*(547). https://doi.org/10.3389/fnhum.2013.00547

Limanowski, J., & Friston, K. (2018). 'Seeing the Dark': Grounding Phenomenal Transparency and Opacity in Precision Estimation for Active Inference. *Frontiers in Psychology, 9*(643). https://doi.org/10.3389/fpsyg.2018.00643

Mackie, J. L. (1973). *Truth, Probability and Paradox: Studies in Philosophical Logic*. Clarendon Press. Retrieved from https://global.oup.com/academic/product/truth-probability-and-paradox-9780198244028?cc=ir&lang=en&

Mackie, J. L. (1977). Dispositions, Grounds, and Causes. *Synthese, 34*, 99–107. https://doi.org/10.1007/978-94-017-1282-8_6

Nagel, T. (1974). What Is It Like to Be A Bat? *The Philosophical Review, 83*(4), 435–450.

Northoff, G. (2014a). *Unlocking the Brain: Volume 1: Coding*. New York: Oxford University Press.

Northoff, G. (2014b). *Unlocking the Brain: Volume 2: Consciousness*. New York: Oxford University Press. https://doi.org/10.1093/acprof:oso/9780199826995.001.0001

Northoff, G. (2018). *The Spontaneous Brain: From the Mind-Body to the World-Brain Problem*. Cambridge, MA: MIT Press. Retrieved from https://mitpress.mit.edu/books/spontaneous-brain

Northoff, G., Tsuchiya, N., & Saigo, H. (2019). Mathematics and the Brain—A Category Theoretic Approach to Go Beyond the Neural Correlates of Consciousness. *BioRxiv*, 674242. https://doi.org/10.1101/674242

O'Brien, G., & Opie, J. (2004). Notes Toward a Structuralist Theory of Mental Representation. In H. Clapin, P. J. Staines, & P. P. Slezak (Eds.), *Representation in Mind: New Approaches to Mental Representation*. Pergamon: Elsevier.

Oizumi, M., Albantakis, L., Tononi, G., Thompson, C., Tononi, G., Tononi, G., ... Koch, C. (2014). From the Phenomenology to the Mechanisms of Consciousness: Integrated Information Theory 3.0. *PLoS Computational Biology, 10*(5), e1003588. https://doi.org/10.1371/journal.pcbi.1003588

Psillos, S. (2002). *Causation and Explanation*. Kingston, ON: McGill-Queen's University Press.

Psillos, S. (2006). The Structure, the Whole Structure, and Nothing But the Structure. *Philosophy of Science, 73*(5), 560–570. https://doi.org/10.1086/518326

Ramstead, M. J. D., Badcock, P. B., & Friston, K. J. (2018). Variational Neuroethology: Answering Further Questions. *Physics of Life Reviews*. https://doi.org/10.1016/j.plrev.2018.01.003

Rosenthal, D. (2012). Higher-Order Awareness, Misrepresentation and Function. *Philosophical Transactions of the Royal Society of London B: Biological Sciences, 367*(1594), 1424–1438.

Salmon, W. C. (1984). *Scientific Explanation and the Causal Structure of the World*. Princeton, NJ: Princeton University Press.

Searle, J. R. (1983). *Intentionality: An Essay in the Philosophy of Mind*. Cambridge: Cambridge University Press.

Searle, J. R. (2013, January 10). Can Information Theory Explain Consciousness? *The New York Review of Books*. Retrieved from http://www.nybooks.com/articles/2013/01/10/can-information-theory-explain-consciousness/

Shagrir, O. (2012). Structural Representations and the Brain. *The British Journal for the Philosophy of Science, 63*(3), 519–545. https://doi.org/10.1093/bjps/axr038

Shannon, C. E., & Weaver, W. (1949). *The Mathematical Theory of Communication*. Urbana, IL: University of Illinois Press.

Shea, N. (2016). Representational Development Need Not Be Explicable-by-Content. In *Fundamental Issues of Artificial Intelligence* (pp. 223–240). Cham: Springer International Publishing. https://doi.org/10.1007/978-3-319-26485-1_14

Swoyer, C. (1991). Structural Representation and Surrogative Reasoning. *Synthese, 87*(3), 449–508. https://doi.org/10.1007/BF00499820

Tononi, G. (2004). An Information Integration Theory of Consciousness. *BMC Neuroscience, 5*(1), 42. https://doi.org/10.1186/1471-2202-5-42

Tononi, G. (2015). Integrated Information Theory. *Scholarpedia, 10*(1), 4164. https://doi.org/10.4249/scholarpedia.4164

Tononi, G. (2017a). Integrated Information Theory of Consciousness: Some Ontological Considerations. In *The Blackwell Companion to Consciousness* (pp. 621–633). Chichester: John Wiley & Sons, Ltd. https://doi.org/10.1002/9781119132363.ch44

Tononi, G. (2017b). The Integrated Information Theory of Consciousness: An Outline. In *The Blackwell Companion to Consciousness* (pp. 243–256). Chichester: John Wiley & Sons, Ltd. https://doi.org/10.1002/9781119132363.ch17

Tononi, G., Boly, M., Massimini, M., & Koch, C. (2016). Integrated Information Theory: From Consciousness to Its Physical Substrate. *Nature Reviews Neuroscience, 17*(7), 450–461. https://doi.org/10.1038/nrn.2016.44

Tononi, G., & Koch, C. (2015). Consciousness: Here, There and Everywhere? *Philosophical Transactions of the Royal Society of London B: Biological Sciences, 370*(1668), 1–18.

Tschacher, W., & Haken, H. (2007). Intentionality in Non-Equilibrium Systems? The Functional Aspects of Self-Organized Pattern Formation. *New Ideas in Psychology, 25*(1), 1–15. https://doi.org/10.1016/J.NEW IDEAPSYCH.2006.09.002

6

Social and Moral Aspects of the Self

6.1   Social and Ethical Aspects

Let me begin with a quick recap. In Chaps. 3 and 4, I spelt out and defended a structural realist theory of selfhood (SRS). The construction of SRS starts from the widely observable fact that there is theoretical diversity between scientific theories of selfhood in psychology and neuroscience. An antirealist may take an instrumentalist attitude towards this theoretical diversity without assuming that any of the theories convey implications about the real structure of the self. However, the theoretical diversity can be a conundrum to the realist who wants to inform their metaphysics of selfhood by the best scientific accounts of the self. This is because the implications of different theories can be inconsistent. Under the circumstances, SRS suggests that it is best to make ontological commitments to the commonality, or the common underpinning structure that lies beneath theoretical diversities and subsumes them. The outcome is a unifying structural realist account of the self. I also addressed the question of the nature of the underpinning

Parts of this chapter are reprinted with the kind permission of Springer Nature.

© The Author(s) 2019
M. D. Beni, *Structuring the Self*, New Directions in Philosophy and Cognitive Science,
https://doi.org/10.1007/978-3-030-31102-5_6

structure to which we can make ontological commitments. In Chap. 4, I showed that at the level of embodied neuronal mechanisms, we can specify the underpinning structure of the self as embodied informational structures or informational mechanisms that are implemented in the brain's cortical midline structures (CMS) and its Default Mode Network (DMN). In Chap. 5, I proceeded to account for some phenomenal aspects of the self—for example, consciousness, self-consciousness, and intentionality—on a structural basis. I showed that it is possible to address the issue of the theoretical diversity of various structural accounts of phenomenal aspects by taking a structuralist strategy, that is, by emphasising the role of the underpinning basic structure of consciousness. As we have already specified basic self-structure and taken care of phenomenal aspects, we can attend to social and moral aspects of the structural self. In this chapter, I shall proceed to develop a structural account of the social and moral aspects of the self. The structural account of social and moral aspects of the self that I will develop in this section is informed by state-of-the-art research in the neuroscience of the Mirror Neuron System (MNS) and the Default Mode Network (DMN). Previously (in Sects. 4.5.1 and 4.5.2 of Chap. 4) I specified the basic structure of the self in terms of the MNS and DMN. Therefore, my account of the social and moral aspects of the structural self meshes nicely with what I said about the basic structure of the selfhood. That is to say, my plan for a synthesis between the social and reflective aspects of the self receives support from experimental research on the dynamical relation between the MNS and the DMN, which respectively realise the patterns of social and reflective aspects of the self. This chapter also provides a broad-brush sketch of a structural account of the moral aspects of the self (or morality of the structural self). I briefly argue that a viable theory of morality must find a middle ground between complete selfishness and extreme selflessness (or alienation). Additionally, a viable theory of morality should not conceive of the self as being disintegrated into affective and cognitive halves. I will argue that a structural account of the moral aspects of the self satisfies both conditions.

6.2   Neurological Bases of Social Cognition

We aim to show how the self is embedded in its social manifold on the basis of the bodily neurological mechanisms that make the self-world relationship possible. Some remarks on the neurological basis of social cognition would be in order. The family of Theories of Grounded Cognition (TGC for short) provides such a neurological basis.

TGC is inspired by the works of French philosopher Merleau-Ponty and the pragmatist John Dewey (Clark, 2016; see Varela, Thompson, & Rosch, 1991). It also draws on psychological theories of J. J. Gibson, Lakoff, Barsalou, and a few others. An interesting overview of the history and main themes of TGC is offered by Glenberg (see Glenberg, 2010 section VARIETIES OF EMBODIMENT). Here, I briefly point out that TGC aims to unify theories of learning, memory, perception, and cognition by underscoring the role of action and goal-directed motor control in guiding cognition and perception. This is in contrast with the classical, amodal accounts of cognition, which presume that interaction of the organism with the environment (or its social network) takes place mainly through abstract representational models. Instead, TGC indicates that the sensorimotor system is only in charge of controlling the organism's movement. The motor control mechanisms underlie the central processes that are involved in both goal-directed action and cognition. Drawing on the resources of evolutionary psychology, TGC emphasises the coevolution of body and behaviour and the interconnection of action, emotion, and cognition. In this vein, it challenges the orthodox accounts of cognition and perception, which assume that representations are processed as amodal abstract data structures (Barsalou, 2008; Glenberg, 2010; Pezzulo et al., 2012).

There are evolutionary reasons for underscoring the role of bodily processes and sensorimotor mechanisms in perception and cognition. For example, it can be remarked that psychological phenomena have to be organised around the role of the action and its bodily characteristics (Glenberg, 2010, pp. 586–587). From an evolutionary perspective, brains aim at guiding interaction with the world, instead of forming abstract models of the world's features. Because bodies mediate this interaction, brains and their cognitive faculties have to coevolve with bodies so as to be able to achieve their evolutionary goal. The nervous system aims to guide action, and since action is also dependent on the body and its capacities, the cognitive system evolves to make embodied cognition possible. There is ample experimental evidence to confirm the plausibility of TGC and its account of the interconnection between actions, emotions, cognition, and the embodiment of the corresponding bodily patterns (Barsalou, 2008; Glenberg, 2010; Grant & Spivey, 2003; Pezzulo et al., 2012). The evidence supports the claim that far from being exclusively the result of abstract symbol processing mechanisms, cognitive phenomena depend upon the sensorimotor brain regions, for instance, premotor cortex, primary motor cortex, cerebellum, and so on, as well as bodily mechanisms that aim to make action possible. TGC provides interesting accounts of social cognition and inter-subjectivity. Let me elaborate.

In Sect. 4.5.2 of Chap. 4, I alluded that TGC includes the study of the Mirror Neuron System (MNS), which accounts for the self's direct and unreflective understanding of other people's mental states, their desires, beliefs, and intentions, and so on. This system can be defined as "a collection of brain regions that are active when we do or experience something ourselves, and also when we observe someone else doing the same thing or having the same experience" (Frith, 2008, p. 2033). The general insight is that when we observe an action, our motor system automatically and unconsciously simulates the performed action, albeit without really replicating the action (Gallese, 2014). This means that the self could simulate and understand the actions of others and their mental state without actually reproducing or executing their actions. In Sect. 4.5.2 of Chap. 4, I also explained that mirror neurons are multimodal motor neurons that populate ventral premotor area F5, area FF, and the inferior parietal lobe of the brains of macaque monkeys. The neurons are active both when the monkeys engage in object-related action or/and when they observe others' (monkeys' or humans') actions (Gallese, Keysers, & Rizzolatti, 2004; Umiltà et al., 2001). These neurons are sensitive to partly concealed actions and motor goals as well as observable actions (Gallese, 2003, p. 522; Umiltà et al., 2001). It is possible to account for social cognition on the basis of the operation of the MNS. Below, I shall unpack this last remark.

Some primates, for example, hominoids, live in complex groups and engage in a series of inter-individual relations. Engaging in inter-individual relations in complex social groups brings about certain problems as well as some evolutionary advantages. The individuals must be able to ascribe mental states to other individuals rather precisely in order to communicate successfully with them. Once the possibility of effective communication is secured, the individuals and species can jointly maximise survival by optimising their knowledge through communication, for instance, by trading off their information about danger and avoiding it with each other, or collaborating in their search for resources of nutrition and exploiting the environment. The MNS is the neuronal mechanism that underpins effective inter-individual communication about the environment in primates (Parr, Waller, & Fugate, 2005). Mechanisms of effective interpersonal communication—for example, facial expressions, self-conscious emotional representations, referential vocalisation, and tracing eye gaze cues—are implemented by mirror neuron structures in prefrontal neuro-cortex, regions which include spindle cells, and von Economo neurons (ibid.). The function of the MNS is almost the same across human brains and other hominoids' brains (such as macaques' brains). In humans, too, the observation of the goal-directed actions of others is associated with the activation of premotor and parietal areas of the human brain (Gallese, Eagle, & Migone, 2007). In humans, too, the MNS contributes to the imitation of simple acts, learning complex unpractised motor acts, and understanding others' actions and intentions. The activity of the Mirror Neuron System could be specified both in new-born children and adults (Gallese, 2003; Wohlschläger, Haggard, Gesierich, & Prinz, 2003). Mindreading or mentalising, understanding affections such as pain or disgust, action understanding, and empathy are part of the general model of social cognition, which is centred on the function of the MNS. The MNS theory provides a neurological explanation of how the observation of actions stirs the unconscious and automatically simulated re-enactment of the same action in the observer. It is possible to build upon this neurological account to explicate highly complicated social phenomena which contribute to our ability to know the world and others. The MNS also enables us to forge social networks and engage in social interactions because mechanisms of social cognition and consciousness could collaborate in realising shared goal-directed actions. Accomplishment of shared actions presumes the possibility of interaction of at least two agents that need to communicate effectively and trust one another, where trust demands the competence to discern honesty from deception on the basis of the other's conscious and unconscious cues (Frith, 2008, p. 2036 ff.; Frith & Frith, 2007). In the same vein, it is possible to rely on the MNS mechanisms to account for high-level social processing which bestows upon the agents their capacity to develop language, improve their institutional learning system, and so on (Frith & Frith, 2007). In the next section, I shall proceed to show how Gallese has incorporated the experimental research on the MNS and social cognition to his structural account of the social self, that is, a self embodied in the manifold of social relations.

6.3   A Relational Theory of Social Self

Vittorio Gallese has drawn on experimental research on the MNS to present a theory of social aspects of the self as being situated in the social web of interpersonal relations. More interestingly, Gallese has developed his account of social aspects of the self along the lines of structuralism (or relationalism, according to his own vocabulary). According to Gallese, being embedded into the web of social relations is an intrinsic feature of the self and "the notion of intersubjectivity is intrinsically related to the notion of the self" (Gallese, 2014, p. 1). This definition of the intrinsic property of the self indicates that the self can be essentially identified on account of its place within the structure of social relations. In this sense, Gallese's theory of Embodied Relational Self (ERS) is committed to structuralism. Relational and structural can be used interchangeably in the context of Gallese's work. ERS is evolved out of TGC (explained earlier in this chapter) and the MNS-based account of social cognition. According to ERS, focusing on the function and the structure of the experiences of the self from the first-person perspective is not enough for providing a viable understanding of the nature of the self. Gallese identifies the self in relation to the webs of social interrelations which could be specified on the basis of operations of the bodily mechanisms and the Mirror Neuron System.

ERS draws on four premises. It presupposes the existence of an action-capable agent with the capacity for ownership; it rests on (and is limited by) the capacities of motor system; it presumes the possibility of inter-connection between perception and goal-directed action; and it presumes that self-related multimodal sensory information about the body is not dislodged from the world it interacts with (Gallese, 2014, p. 2). ERS also underscores the role of the motor system in constituting the basic sense of the self and shaping its perception and pre-reflective conception of others, given that others have motor capacities and experiences similar to ours. According to Gallese, "When exposed to others' expressive behaviours, reactions and inclinations, we simultaneously experience their goal-directedness and intentional character, as we experience ourselves as the agents of our actions, the subjects of our affects, feelings and emotions, the owners of our thoughts, fantasies, imaginations and dreams" (Gallese, 2014, p. 5). This basic understanding provides the foundation of an epistemically deflationary account of other minds.

ERS holds that the self is generated in a shared multidimensional interpersonal space (Gallese, 2003; Gallese et al., 2007). As I have argued, ERS is a relational theory. It is based on the neurologically informed account of the self-others relationship and it assumes that the self and the social manifold that situates it are interdependent. The emphasis on the self-others relationship indicates that the constitution of the self is intrin-sically dependent upon its role in the network of social relations. Thus, Gallese defines the self in terms of its relational properties, which are defined by virtue of the self's place in the manifold of interpersonal rela-tions. This definition is thoroughly structural. The social aspect of the self and its relation to the social structure is an intrinsic property of the self, and thus the self is essentially a relational (or structural) entity according to Gallese. Before proceeding further, I have to clarify a point.

There can be (and indeed are) various ways of developing a plausible theory of social cognition without making the account of social cognition dependent on the mechanism of the MNS. For example, Ratcliffe devel-oped a theory of social cognition (in terms of understanding another

person) based on Strawson's orthodox notion of "person" as an inscruta-
ble basic unit. Ratcliffe argues that accounting for social cognition in
terms of the mechanisms of the MNS (or probably any other sub-personal
mechanisms of simulation) is not enough for providing an account of
social cognition, because we must establish the notion of a person as the
bearer of the mental states first, that is, before presenting a theory of
social cognition on the basis of simulating others' mental states (Ratcliffe,
2013, p. 224).

I have already raised issues with the substantivalist theory of personhood (in Chaps. 1 and 3). Arguably, even if a theory of social cognition
mandates explaining the experience of the other as a substantive entity,
this does not mean that the persons are ontologically basic units. In a
charitable reading, I assume that Ratcliffe's theory does not need to
restore the substantivalist conception of the self (if it does, I challenge his
theory on grounds mentioned in Chaps. 1 and 3). Rather, it seems that
Ratcliffe intends to develop an embodied account of social cognition
according to which the recognition of a person consists in a "bodily feeling that is at the same time (a) an acknowledgement of the other person
as a locus of experience and activity distinct from oneself and (b) a change
in how one experiences the world" (Ratcliffe, 2013, p. 225). He develops
this account to replace the account of understanding the other in terms
of simulation with direct bodily experience of the other. In this vein,
Ratcliffe leans towards theories of radical embodiment and enactivism
(e.g., Gallagher, 2014; Hutto, 2015). This means that Ratcliff dispenses
with mindreading and simulation (say, as formed within the MNS) and
makes the account of social cognition dependent on direct experience of
the other (unlike some advocates of the embodied account, he still gives
the experience of the person priority over interpersonal interactions or
the shared world). However, I do not think the emphasis on the embodied aspects of social cognition needs to undermine the role of the MNS
and mechanisms of simulation. This assertion is in line with my elaborations on the moderate embodied approach of this book (see Sects. 4.6
and 4.7 in Chap. 4). The moderate version of embodiment that I advocate can motivate a view of social cognition that involves both "mindreading" and "embodied simulation". This is because it enriches
embodiment with elements from the theory of mind and simulation

theory and accordingly enables me to avoid taking a stand on some of the key debates surrounding the nature of social cognition. Below, I shall explain my take on the moderate embodied view.

The embodied, environmental factors are too important to be ignored in a viable theory of selfhood (and its aspects). However, eliminating representations and simulations from the account of cognition would result in a form of theoretical poverty that may damage the plausibility of the account of the relationship between the agent and its environment. If agents want to interact with their environment, they must use an intermediary, in terms of models or representations, even if we assume that the account of model making and representation is embodied. Also, the direct embodied psychological account of the organism-environment relationship, even when presentable at all, does not identify with a metaphysical or epistemological theory of direct perception or direct realism (Drayson, 2018).

On the same subject, my moderate approach to embodiment has its home in neuroscience. This means that basic aspects of the self are embodied in the mechanisms of the brain and cognitive system, more or less in the sense that is unfolded by Gallese (2017). As I explained in Chap. 4 (Sects. 4.6 and 4.7), the brain is not secluded from the world and the embodied informational structure of the self is coupled with (or weaved into) the embodied informational structure of the ecosystem and society. So, although I give priority to the embodied structures of the brain and cognitive system (as the orthodox account gives priority to the concept of person), I consent that the embodied self-structure latches onto the vast infrastructure of the environment and the web of social relations to forge a comprehensive coupled structure (Beni, 2019). This is in line with Gallese's insight, according to which "cognitive neuroscience should resist the solipsistic 'brain-in-a-vat'-like attitude purported by classic cognitive science and study how situated brain-bodies map their mutual relations and their interactions with the physical world" (Gallese, 2017, p. 311).

To recap, there are scientifically informed theories of social cognition that do not depend on the mechanisms of MNS, embodied simulation, mindreading, and so on. I have already alluded to some of these theories (Gallagher & Allen, 2016; Hutto, 2015; Ratcliffe, 2013). I advocate a moderate form of embodiment that allows for representations of other people's mental states as simulations by the MNS mechanisms. Accordingly, the theory of social aspects of the self that I will develop in the remainder of the book remains loyal to the general insight, according to which the basic structure of the self is underpinned by the coupled CMS-MNS system (also see Chaps. 3 and 4). In the next section, I embark on incorporating Gallese's account of social self into SRS.

6.4   Incorporating Embodied Relational Self into SRS

6.4.1   A Philosophical Synthesis

Gallese's aim is to construct a structural theory of the social aspects of the self. ERS accomplishes this objective beautifully. However, the self includes other aspects, in addition to social ones. Reflective and phenomenal aspects are amongst other aspects that have been discussed in this book. Theories of embodied simulation and Mirror Neuron System are mainly about the relation between the self—as a node in the shared manifold of the interpersonal relations—and others who feature in the shared manifold. Although the MNS is capable of processing self-related information, ERS does not use the MNS to explain non-social self-related aspects of the self. That is to say, ERS mainly relies on the MNS to characterise the connective links through which the self latches onto the web of the interpersonal manifold. It does not aim to account for reflective or phenomenal aspects of the self. It does not explain how the reflective aspects are realised by shared patterns of the processing of the Mirror Neuron System. This is not necessarily a shortcoming of Gallese's account, which aims to specify the connection between the self and social cognition. However, a comprehensive account of the structural self, for instance, SRS, needs to consider both social and reflective aspects. Thus far, in presenting SRS, I have elaborated on the basic structure of the self and its phenomenal aspects. I also relied on Free Energy Principle (FEP) to flesh out the biological and ecological plausibility of the self-structure which accommodates the reflective and phenomenal aspects. Nevertheless, SRS still lacks a structural account of the social aspects of the self. That requisite account could be delivered via a synthesis with Gallese's ERS.

SRS, as being developed in the last three chapters of this book, already includes a structural account of the reflective and phenomenal aspects of the self. What it lacks is a neurologically informed structural account of the social aspects of the self. ERS is informed by a neuroscientific account of simulating mechanisms that underlie understanding the actions and intentions of others from the second-person perspective. However, ERS is almost silent about the reflective properties of the self from the first-person perspective. Here, I argue that it is possible to incorporate ERS into SRS, which includes an extensive account of reflective and phenomenal aspects of the self. Although ERS does not take the reflective aspects of the self (from the first-person perspective) into account, when incorporated into SRS, it can easily amend the shortcomings of SRS which leaves out the social aspects of the self. The outcome of the synthesis is a comprehensive version of SRS which can account for all of the salient aspects of selfhood, for when being supplied with ERS, SRS emerges as an almost comprehensive structural account of the self. In addition, as I shall explain immediately, there is a nice scientific footing for supporting the attempt at incorporating ERS into SRS.

Both SRS and ERS are scientifically informed theories. They build upon progressive scientific theories about CMS, FEP, and the MNS to inform their respective accounts of reflective, phenomenal, and social aspects of the self. We embark on incorporating ERS into SRS, so as to achieve a comprehensive structural account of the self. But to consolidate this goal, we must be able to show that there is a scientific basis for forging a connection between the respective neural underpinnings of SRS and ERS. As it happens, the psychological literature contains a number of interesting suggestions that relate the respective functions of CMS and the Mirror Neuron System (Cisek, 2007; Cisek & Kalaska, 2010; Giambra, 1995; Keysers & Gazzola, 2007; Mitchell, Macrae, & Banaji, 2006; Rizzolatti & Luppino, 2001; Uddin, Iacoboni, Lange, & Keenan, 2007). In Sect. 4.5.2 of Chap. 4, I have referred to parts of this literature to show how mechanisms of CMS and the MNS are coupled with one another. That is to say, I referred to the literature to highlight the point that both

CMS and the MNS are inclusively and dynamically engaged in processing information about the reflective aspects and embodied-social aspects in oneself and others. Both systems are engaged in linking cognition and action and emotions in oneself with their counterparts in others. The fact that the information processing activities of CMS and the MNS weave together provides a neurological basis for consolidating the assumption of a close connection between SRS—which is underpinned by CMS's activity—and ERS—which is based on the MNS's activity. As I have argued before (Beni, 2019), this scientific basis consolidates our attempt at incorporating ERS into the outline of a comprehensive structural realist theory of selfhood. In the next sub-section, I will furnish fresh evidence to demonstrate that there is meaningful anatomical and informational interconnection between the brain regions that process self-related information and parts that are concerned with social cognition.

6.4.2   Experimental Support for the Integration

As I have remarked in Chap. 4, CMS is a part of the brain's Default Mode Network (DMN). Mechanisms of the DMN are active when the brain is not engaged in processing external stimuli but is in the resting state. In Chaps. 4 and 5, I explained that CMS, the DMN, and the resting state activity realise the basic structure of the reflective and phenomenal aspects of the self. Also, in Sect. 4.5.2, I explained that there is a meaningful relationship between the neural underpinning of reflective aspects of the self in CMS and the DMN and the neuronal underpinning of imitation and action observation in the MNS (also see Uddin et al., 2007). This neurological entwinement can consolidate the attempt at incorporating ERS into SRS. Here, I will refer to new experimental research, which shows how the interconnection between the DMN and the MNS contributes to the formation of a bigger network that underpins mechanisms of cognition of mental and physical aspects of oneself and others. In a recent experimental study, Molnar-Szakacs and Uddin proposed that:

> many of the same neural systems are engaged for self- and other-understanding. Thus, having privileged access to our own physical and

mental states allows us to gain insight into others' physical and mental states through the processes of embodiment and mentalizing. These cognitive processes are supported at the neural level by two large-scale, interacting networks—the MNS and the DMN, respectively. A more in-depth understanding of the functionally relevant nodes of each network, and the interactions between them, will help us advance toward a more complete theory of self-representation. By bringing together recent work on the fractionation of these complex networks, we aim to contribute to a more complete understanding of the self. (Molnar-Szakacs & Uddin, 2013, p. 3)

According to Molnar-Szakacs, neural mechanisms that underpin cognition of mental aspects of oneself and others are realised by parts of the DMN (e.g., MPFC node), whereas embodied simulation processes that enable representation of embodied and physical aspects of others are implemented in the MNS, which allows us to simulate actions of others by embodying them. Accordingly, it can be argued that the DMN engages in mentalising, that is, the inference-based act of ascribing reflective states to oneself or others, whereas the MNS is in the business of embodied simulation of others' actions. However, during self-related information processing, cognition of mental and embodied aspects of oneself and others do not take place separately in different brain regions. Instead, as experimental studies indicate, the processing of self-related information (at least when based on communicative gestures) is enabled by a dynamical trade-off of information between the DMN and the MNS, where the MNS provides feedback information to the DMN's work (Molnar-Szakacs & Uddin, 2013; Sandrone, 2013; Schippers & Keysers, 2011). Such studies indicate that the DMN and the MNS are coupled through dynamical interaction when the brain engages self-related processing within the context of its social relations, say, in tasks such as recognising one's face and distinguishing it from the face of others, or self-voice recognition. Therefore, there is a meaningful and significant neurological relationship between the MNS and DMN.

Let us recap. There is the interrelationship between functions and information processing mechanisms in the MNS and DMN. It has been presumed that the MNS is in the business of social cognition. However, experimental studies indicate that some sub-regions of the DMN—for example, the medial prefrontal cortex (MPFC)—are engaged in social cognition, empathy, mentalising, and emotion (Li, Mai, & Liu, 2014). Li et al.'s indicate that the DMN regions are not exclusively concerned with the cognition of reflective and personal aspects of the self and can collaborate with the MNS in realising the social aspects of selfhood. Indeed in their construal of similar experimental evidence, Molnar-Szakacs and Uddin (op. cit.) suggest that imaging studies of the anatomical functions of the brain support the assumption of a dynamical interconnection between the DMN and MNS. The DMN and the MNS are integrated functionally and anatomically. Further evidence suggests that disorders in mechanisms of self-related information processing can be traced back to deficient networks in either the DMN or the MNS or both. Thus, there are both positive and negative pieces of experimental evidence to support the assumption of the dynamical interconnection or even integration of the DMN and MNS. Our understanding of our selves in their wholesome glory, as the selves that manifest both reflective and social aspects, is not a result of isolated mechanisms of information processing in independent regions. It is generated by the unified informational networks in the MNS and DMN. This is in line with further experimental studies that demonstrate that anatomical regions that implement the DMN could be heterogeneous, and in addition to CMS regions—for instance, MPFC and PCC—they may include regions that overlap with the MNS—for example, including the Inferior Frontal Gyrus (IFG) and AI—(Leech, Kamourieh, Beckmann, & Sharp, 2011). Thus, there is connectivity between the DMN and MNS sub-regions, and both systems are involved in realising the integrated structure of the self which subsumes personal and impersonal cognition and affection. The evidence ensures that ERS can be incorporated into SRS on the basis of viable theoretical and experimental grounds. According to Molnar-Szakacs and Uddin's construal of these studies,

The DMN and MNS may interact at certain "rich-club" nodes, including the AI and the PCC. Through this interaction, embodied simulation-based representations serve to scaffold mentalizing-based representations. These representations allow the brain to construct a dynamic self, continuous through time, and able to plan for the future. (Molnar-Szakacs & Uddin, 2013, p. 8)

Molnar-Szakacs and Uddin's concluding remark supports our attempt at incorporating ERS to SRS. The endeavour for incorporating the structural account of social aspects of the self into a comprehensive structural account which also includes reflective and phenomenal aspects is not merely a manifestation of philosophical wishful thinking. There are reliable theoretical and experimental reasons to support the assumption of the relationship between the DMN (or CMS) and MNS. Experimental results support the integrative assumption, and once the goal of incorporation is carried on such solid grounds, we can achieve a comprehensive and unified account of the structural self which subsumes all reflective, phenomenal, ecological, and social aspects of the selfhood. Our long endeavour of constructing SRS seems to be accomplished now.

6.5 Some General Remarks on the Moral Aspect of Structural Self

Ethical issues are too important and intriguing to be treated in passing at the end of a chapter that is mainly concerned with the social aspects of the self. However, because ethics is concerned with norms of evaluation and action in a social context, and social and moral aspects are usually connected with one another, I intend to briefly consider the issue of moral aspects of the structural self in this section. Because ethical questions bear on the normative aspects of the self-others relations in a social context, what I will say about the moral aspects of the self will be in harmony with what I have already said about the social aspects of the self. In this section, I make some general remarks about the neurological basis of moral cognition and affection and elaborate on their relationship with the underpinning structure of the self, a structure which also includes social aspects of the self. After elaborating on the neurological underpinning of the self's morality, I will draw on some classical theories of normative ethics to show that there are viable philosophical insights that line up with the structural realist theory of moral aspects of the self.

6.5.1   The Neurological Basis of the Self's Moral Cognition

The same neural structures (i.e., the DMN and MNS) that jointly underpin the reflective and social aspects of the self-structure also underpin mechanisms of moral cognition and affection (Molnar-Szakacs & Uddin, 2013, p. 7). The DMN and MNS also underpin the self's capacity to issue moral judgements in the context of the self-others relationship, and the basic neuronal structure of the self has the capacity to realise the moral aspects of the self.

Let us begin with considering the role of the DMN in the formation of moral judgement. In their search for the origins of social understanding in the brain, Li et al. (2014) have suggested that the DMN and resting state activity are involved in mechanisms of social understanding of others, the theory of mind, empathy, and moral judgements, as well as the processing of self-related information. To flesh out their point, Li et al. have reviewed neuroimaging research on resting state activity of the brains of subjects with normal functional connectivity in their DMNs as well as research that studies the brains of abnormal subjects with defective functional connectivity in the DMN or abnormal resting state activity, for instance, due to cocaine dependence. The experimental studies suggest that webs of connectivity within the DMN—for example, MPFC-PCC and vMPFC-TPJ—are involved in dealing with moral dilemmas and surveying moral scenarios (Li et al., 2014, p. 4). It is also possible to find negative correlation between damage in the functional connectivity between the DMN regions—for instance, between the medial frontal cortex (aDMN) and posterior brain areas (pDMN)—on the one hand and lack of moral sensitivity, say in psychopathic persons with documented criminal histories (Li et al., 2014, pp. 8–9). Through positive and negative arguments, Li et al. have established the integrative coherence of mechanisms of social cognition, empathy, mentalising, and morality at the level of the DMN and resting state activity. There are also other fMRI studies which ground the competence for evaluation of actions of oneself and others in the resting state activity of the DMN (e.g., Jung et al., 2016).

The DMN and resting state activity underpin the basic structure of the self. The same basic structure can also exhibit the moral aspects of self-hood. However, it is important to note that moral judgements are not exclusively underpinned by the DMN and resting state activity. On the same subject, note that in their account of the involvement of the DMN in moral judgements, Li et al. (2014) also drew attention to functional connectivity between the DMN and other regions of the brain. They argue that there is observable connectivity between the DMN and other regions—for example, vMPFC-amygdala and ACC-thalami—that are engaged in making moral judgements. The DMN is therefore connected to other brain regions that underpin mechanisms of moral judgement. This point, about the overlap of several brain circuits—for instance, parietal and temporal structures, the insular cortex, and subcortical structures—is also explicated in Pascual et al.'s account of emotionally driven moral decisions (Pascual, Rodrigues, & Gallardo-Pujol, 2013). Pascual et al. also remark that moral decisions are based on emotional reactions as well as cognitive deliberations (Pascual et al., 2013, p. 2). However, this does not need to indicate that moral decisions are split into emotional vs. cognitive decisions, and there are reasons for thinking that the cognitive and emotional mechanisms that contribute to the formation of moral judgements are integrated. To substantiate the integration between the cognitive and emotional mechanisms of moral judgement, we can show that moral judgements are the result of the dynamical relationship between the brain structures that are engaged in cognitive reflection and those that underpin emotions, empathy, and social cognition. In other words, the traced dynamical relation between the DMN and the MNS can also vouchsafe the integration of cognitive and affective mechanisms of moral decision-making. Pascual et al. support this proposal with experimental findings. They suggest that the dynamical connectivity between various circuits and brain regions, for example, frontal lobe, parietal lobe, and limbic lobe, subsumes the integration of emotional and cognitive mechanisms of moral decision-making. The dynamical interaction allows for contextualisation of cognitive and emotional information processing mechanisms that contribute to the formation of moral judgements in different situations and in accordance with social norms and values (Pascual et al., 2013, p. 2). The MNS is among the systems whose dynamical relationship with the DMN underlies moral judgements. The point about the existence of structural and functional relation between the DMN and MNS has been emphasised before in this chapter (with reference to Molnar-Szakacs & Uddin, 2013). As I will explain shortly, the MNS and the DMN collaborate in the formation of moral judgements too.

I have already elaborated on the role of the DMN in the generation of moral judgements. As I have explained before in this chapter, the MNS underpins social cognition, understanding others, and affections. Molnar-Szakacs (2011) explains that by inducing empathy and the desire to alleviate other people's suffering, the MNS contributes to providing the neurological substrate of the evolution of social norms and morality. The mechanisms of empathy find a central role here. The MNS underpins the mechanisms of social cognition, which include simulation of others' feelings and emotions as our own and invoking compassion through interaction with others who possess similar mechanisms of cognition and action. I unfold this point immediately by referring to the case of understanding facial expressions. It has been argued that by simulating the relevant activation patterns—for instance, in the premotor face area, the posterior IFG, and the superior temporal sulcus—the MNS enables the mechanisms of understanding other people's facial expressions (Molnar-Szakacs, 2011, p. 78). Understanding facial expressions of others contributes to reflecting the emotional states of other people in one's mind. This is because the simulation of other people's emotional states triggers mechanisms of empathic response, that is, the desire to alleviate their suffering when one discerns emotional states of others as states of pain or suffering. Thus, empathic responses are based on the MNS's pre-cognitive and pre-reflective mechanisms of simulation of others. It is worth mentioning that, in addition to the empathic responses that are associated with simulating others' emotions without reflection, there are cognitive mechanisms of empathic responses. The latter kind of empathic responses is based on the capacity to actively think about other people's actions and emotions. Such cognitive mechanisms are embodied in the limbic system and are associated with higher-order cognitive faculties. Molnar-Szakacs mentions neuroimaging evidence that confirms the distinction between pre-reflective mechanisms of emphatic response embodied by the MNS and cognitive mechanisms that are implemented in the limbic system. It has also been argued that pre-reflective mechanisms may provide inputs to more sophisticated cognitive mechanisms of empathy (Molnar-Szakacs, 2011, pp. 79–80). The interplay between affective and cognitive mechanisms of empathy gives rise to sophisticated forms of social and moral judgement. While observing another person's pain usually induces pain in us—owing to the MNS's pre-reflective simulations—the perceived fairness/unfairness of the observed others modifies the amount of the pain simulated by the observer. Molnar-Szakacs mentions neuroimaging studies which confirm these assumptions. There are studies which indicate that if we think that the person who is in pain is unfair, our empathy-related responses to the observation of that person's pain will be significantly reduced (Molnar-Szakacs, 2011, pp. 79–80; Singer et al., 2006). Note that judgement of a person's fairness/unfairness requires sophisticated cognitive processing of not only observable behaviour of the person but also various cultural and social elements. Such sophisticated normative judgements that are modified by one's conception of cultural and social accounts of the one-other relationship are based on the dynamical relationship between unreflective and cognitive mechanisms of empathic response at the neurological level. These mechanisms are embodied in extensive neuronal regions which include the limbic system as well as the MNS. This is in line with what I said earlier in this chapter about the collaboration of the MNS and other brain structures, for example, CMS, DMN in underpinning the self-structure at a neurological level. This adds up to the conclusion that the same constellation of neurological structures that underpins the basic structure of the reflective and social aspects of the self also accommodate the moral aspects of the self, without invoking anything similar to the orthodox self-substance as the bearer of moral properties.

Let us recap. The basic structure of the self exhibits moral and social aspects of the self as well as phenomenal and reflective aspects. We have accomplished a comprehensive structural account of the self eventually. Our account of the moral aspects of the self, as unfolded in this section, is in harmony with the general structural realist theory of the self that is developed in this book. The account of the moral aspects of the self receives support from experimental research on the neurobiological substrate of moral judgements. However, the remaining question is that the experimental support aside, does it really make sense to ascribe moral faculties to anything but the orthodox substantivalist self? To be clear, it might be assumed that only orthodox selves or persons could be the bearers of moral properties, and eliminating the orthodox self from ontology would damage morality because there remains no one to shoulder the responsibility of moral decisions. In the next section, I argue that it is not only possible to ascribe moral aspects to the structural self, but it is actually the best way to contrive the possibility of accounting for the moral aspects of the self, by striking a fine balance between the two extremes of egotism and alienation. Caring only for oneself and acting solely on the basis of self-interest push one towards egotism, whereas disregarding one's well-being and caring only for others may result in various forms of alienation (from oneself and even from the moral norms). In the next section, I will draw on Peter Railton's (1984) paper on the demands of morality to explain that SRS provides a golden mean between egotism and alienation, and thereby vouchsafe the possibility of moral judgement.

6.5.2   From a *Somewhat* Impersonal Point of View

There is a consensus that morality demands taking an impersonal universalised point of view. Someone who only cares about herself cannot engage in moral action. The point can be easily clarified via an example from a classic British TV show, *Blackadder*. In episode five of season four of the show, Edmund Blackadder, a British army captain, gets involved in a romantic relationship with nurse Mary. The following is extracted from a dialogue between Mary and Edmund;

Mary: Tell me, Edmund: Do you have someone special in your life?
Edmund: Well, yes, as a matter of fact, I do.
Mary: Who?
Edmund: Me.
Mary: No, I mean someone you love, cherish and want to keep safe from all the
            horror and the hurt.
Edmund: Erm … Still me, really.

What is maverick in Edmund's honest reply is that he does not care a jot for the well-being and happiness of anyone except himself. This expressed egotism does not make Edmund immoral but amoral, that is, someone who is not able to behave morally or does not care to do so. One must outgrow their personal concerns in order to care for the other's well-being, in a way that is demanded by morality. One has to suppress their selfishness and egotism to become competent to make moral judgements and engage in moral action. Banishing the orthodox self from the ontology contributes to decreasing the amount of selfishness and egotism of one's self, and thus it facilitates moral behaviour. This indicates that bypassing orthodox substantivalism does not necessarily damage morality but supplies it. However, it remains true that in a world that is totally devoid of selves, moral judgements and evaluations can hardly be accommodated. Below, I shall flesh out this point with an eye to Peter Railton's account of the relationship between morality and alienation.

In his classic "Alienation, Consequentialism, and the Demands of Morality", Railton (1984) endeavours to defend a version of consequentialism that is concerned with the role of personal considerations as well as impersonal and universal perspectives in the fine-tuning of moral judgements. His paper is interesting from our perspective because it considers the integration between affective and cognitive aspects of the self as well as the relation between personal and impersonal factors in the formation of moral judgements. While Railton's argument is developed in the context of consequentialist ethics, as he points out, his views could be applied to other moral perspectives, for instance, deontological ethics, mutatis mutandis. The important point in Railton's paper is that he emphasises the moral significance of the self's care for itself, as well as the compatibility between its affective and rational faculties.

To flesh out his point, Railton offers some examples from the morality of people who, unlike my earlier example of Edmund Blackadder, engage in completely universalised and impersonal moral behaviours. John, for example, is a husband that takes care of his wife's needs with great sensitivity and selflessness. He accounts for his behaviour by saying "I get a lot of satisfaction out of it. Just think how awful marriage would be, or life itself, if people didn't take special care of the ones they love" (Railton, 1984, p. 135). Although John is a paragon of a morally aware husband, he takes a distant, impersonal attitude toward his wife. His attitude is devoid of personal affections, and his rational interest in his wife's well-being takes a certain alienated form, which is a manifestation of the estrangement of the rationality that underpins John's view about how marriages should be free from affections, when the person engage in the married life purely on the basis of duties and responsibilities, instead of any kind of feelings (Railton, 1984, p. 137). Given that, "to have a morality is to make [impersonal] normative judgments from a moral point of view and be guided by them" (ibid., p. 138), John is not neglecting his moral duties by taking an impersonal objective attitude towards his wife's well-being. Rather, the problem is that this alienated attitude may eventually estrange him from his wife, and in the long run, from morality itself, for morality cannot survive after the demise of all non-moral personal factors. This is because, as Railton argues, there are non-moral factors which possess internal value, for example, autonomy and solidity as well as the happiness and well-being of the person. Here, the general insight is that moral goals cannot be pursued as goals external to a person's own perspective and concerns and regardless of what is valuable to the person herself. Morality demands not only a selfless, impersonal perspective but also a personal perspective that incorporates meaning and purposefulness in our moral lives. A totally selfless and impartial morality would be alienated from normal human life and could eventually lead to disregard for morality itself, as something irrelevant to a person's life and well-being. In this context, Railton suggests that instead of removing personal perspectives from moral judgements we may retain the persons and avoid the calamities that are associated with alienation. However, in order to avoid selfishness and egotism, a viable theory of morality may begin with persons not as pre-social atomic individuals—who only pursue their maximal self-advantage—but as individuals that are situated in the structure of social and cultural relations. His view, which will be quoted presently, is evidently consistent with structuralism about the self;

As a start, let us begin with individuals situated in society, complete with identities, commitments, and social relations. What are the ingredients of such identities, commitments, and relations? When one studies relationships of deep commitment—of parent to child, or wife to husband—at close range, it becomes artificial to impose a dichotomy between what is done for the self and what is done for the other. We cannot decompose such relationships into a vector of self-concern and a vector of other-concern, even though concern for the self and the other are both present. [...]. These reference points do not all fall within the circle of intimate relationships, either. Among the most important constituents of identities are social, cultural, or religious ties [...]. Our identities exist in relational, not absolute space, and except as they are fixed by reference points in others, in society, in culture, or in some larger constellation still, they are not fixed at all. (Railton, 1984, pp. 165–167)

As the quotation indicates, Railton suggests dissolving the dilemma—about how to find a middle ground between egotism and alienation—by retaining persons, but instead of conceiving of them as individuals who only serve their self-interests and self-advantage, he defines them as units that are situated in the structure of social and cultural relations. Evidently, Railton's reconstruction of the consequentialist view is compatible with the structural realist theory of the self that is developed in this book. I think this view—which is based on striking a balance between egotism and alienation—can be stated in terms of other theories of normative ethics that are not openly committed to moral egotism. Instead of conceiving of the self as the orthodox substance or an atomic unit with personal properties, Railton reconceptualises the self in terms of the interrelations within the webs of social and cultural structures.

Also, note that his view does not portray the self as divided into cognitive and affective halves. Disregard for one's affections pushes one towards alienation, but, as Railton explains, to be engaged in the web of interpersonal relations, one does not need to give her own affections. What I said in Sect. 6.5.1 about the integration of affective and cognitive neurological mechanisms of empathic response according to moral neuroscience supports Railton's view on unity of affections and cognition of a person in generating well-proportioned moral judgements. On the same subject, Railton defines selves/persons as units that are situated in the structure of social and cultural relations, and this structure seems like yet another structure in which the self is situated. This take on Railton's theory is in line with the moderate embodied approach that I advocate in this book.

Although the self-structure includes some representations and embodied simulation, it is coupled with or situated in the informational structure of social relations. The self-structure is extended from the neuronal regions in the brain and sensorimotor system to the ecological and social context in which the self is situated. Further research may want to focus on developing a structural realist account of the cultural and social infrastructure that encompasses the self-structure and the relation between these two kinds of structures.

6.6   Concluding Remarks

In this chapter, I accounted for the social and moral aspects of the structural self. I endeavoured to show how the structural realist theory of the self reconciles reflective aspects of the self with the social aspects, namely by drawing on the dynamical interrelations between the MNS and the DMN. I also showed how the underpinning neurological structure can be conjured to account for the integration of cognitive and affective mechanisms of empathic response and moral decision as well as the moral aspects of the self in general. In the final section, I showed that not only is SRS compatible with some classical theories of normative ethics but also this scientifically informed account could help us to bypass some of the problems of traditional theories of moral judgement, for instance, the problem of alienation and the problem of affection-cognition disintegration. Below, I shall unfold the main point of this chapter against the general outline of my enterprise in this book.

In this book, I endeavoured to find a viable philosophical alternative to the orthodox substantivalist view of the self. The alternative was supposed to subsume the vagaries of scientific accounts that not only diverge from the substantivalist theory of the self, but also come with inconsistent pictures of the self (e.g., in pluralism vs. eliminativism case). Within this context, I have suggested that we may make ontological commitments to the basic informational structure that lies beneath the theoretical diversity of scientific theories of aspects and elements of the self in cognitive psychology and computational neuroscience. I have also argued that the basic structure of the self could be specified in different ways. It could be specified at an abstract level within the information-theoretic framework of the Free Energy Principle theory. It is also possible to specify the structure at an embodied level, in terms of the coupled mechanisms of the MNS and CMSs, which underpin the informational structure of reflective and social aspects of the self. I accounted for the relationship between the abstract and embodied structures on various occasions (e.g., Sects. 4.7 and 5.5 of Chaps. 4 and 5). In this chapter, I argued that the same informational structure that is embodied in the coupled mechanisms of CMS and MNS underpins the basic structure of moral aspects of the self. The conclusion that I like to draw from this statement is that morality is entwined with the basic underpinning structure of the self and is inseparable from it. This is because the same system (the coupled CMS-MNS system) that underpins the basic structure of the self also embodies the self's capacity to issue moral judgements in the context of the self-others relationship. What better proof could we have to conclude that the capacity to make moral judgements is inseparable from the basic structure of the self? I conclude that selves are constitutively moral beings.

Note that in presenting the self as an essentially moral entity, I did not revert to the substantivalist paradigm. I specified the infrastructure that includes the basic structure of the self (and the self-others relationship) as well as the mechanisms of moral judgement. We do not need a substantial self as the bearer of the burden of moral judgements. The burden lies on the basic structure of the self into which the moral essence of the self is entwined. So, we provide a viable account of important aspects of the self (moral and social aspects) without reverting to substantivalism.

The structuralist account of the social and moral aspects of the self that I have developed in this chapter is strongly naturalistic, in the sense that it is informed by recent breakthroughs in the computational cognitive science of the MNS and CMS. I do not claim that the scientific accounts provide the only viable picture of the nature of the self and its various aspects, moral and social aspects included. Actually, I draw on a classical theory of morality (by Railton) to show that the scientific and philosophical speculations are best when developed jointly, in the context of a meaningful and open-minded dialogue. That being said, as I stated at the beginning, the goal of this book is to present a viable philosophical account of the self, that is, a philosophical view that could provide a basis for unfolding the intellectual implications of the scientific theories of the field, without domesticating them to stagnant metaphysical intuitions. I think, in presenting a well-founded philosophical theory of the self that does not yield to substantivalism, I have delivered on that promise fully now. The structural realist theory of the self (SRS) is well founded in the sense that it is based on venerable precedents in the history of philosophy (see my references to works of Hume and Kant in Chap. 1). The theory also passed its test of plausibility in addressing the quandaries of the recent philosophy of many-particles physics. In this book, I showed that the same structuralist strategy could be used to overcome the confusing state of diversity in scientific accounts of the self and its various aspects. The result is a scientifically informed and philosophically well-founded alternative to the substantivalist theory of the self.

References

Barsalou, L. W. (2008). Grounded Cognition. *Annual Review of Psychology,* *59*(1), 617–645. https://doi.org/10.1146/annurev.psych.59.103006.093639

Beni, M. D. (2019). An Outline of a Unified Theory of the Relational Self: Grounding the Self in the Manifold of Interpersonal Relations. *Phenomenology and the Cognitive Sciences, 18*(3), 473–491. https://doi. org/10.1007/s11097-018-9587-6

Cisek, P. (2007). Cortical Mechanisms of Action Selection: The Affordance Competition Hypothesis. *Philosophical Transactions of the Royal Society of London. Series B, Biological Sciences, 362*(1485), 1585–1599. https://doi. org/10.1098/rstb.2007.2054

Cisek, P., & Kalaska, J. F. (2010). Neural Mechanisms for Interacting with a World Full of Action Choices. *Annual Review of Neuroscience, 33*(1), 269–298. https://doi.org/10.1146/annurev.neuro.051508.135409

Clark, A. (2016). Busting Out: Predictive Brains, Embodied Minds, and the Puzzle of the Evidentiary Veil. *Noûs.* https://doi.org/10.1111/nous.12140

Drayson, Z. (2018). Direct Perception and the Predictive Mind. *Philosophical Studies, 175*(12), 3145–3164. https://doi.org/10.1007/s11098-017-0999-x

Frith, C. D. (2008). Social Cognition. *Philosophical Transactions of the Royal Society of London. Series B, Biological Sciences, 363*(1499), 2033–2039. https://doi.org/10.1098/rstb.2008.0005

Frith, C. D., & Frith, U. (2007). Social Cognition in Humans. *Current Biology,* *17*(16), R724–R732. https://doi.org/10.1016/J.CUB.2007.05.068

Gallagher, S. (2014). Pragmatic Interventions into Enactive and Extended Conceptions of Cognition. *Philosophical Issues, 24*(1), 110–126. https://doi.org/10.1111/phis.12027

Gallagher, S., & Allen, M. (2016). Active Inference, Enactivism and the Hermeneutics of Social Cognition. *Synthese,* 1–22. https://doi.org/10.1007/s11229-016-1269-8

Gallese, V. (2003). The Manifold Nature of Interpersonal Relations: The Quest for a Common Mechanism. *Philosophical Transactions of the Royal Society of London B: Biological Sciences, 358*(1431). Retrieved from http://rstb.royalsocietypublishing.org/content/358/1431/517.short

Gallese, V. (2014). Bodily Selves in Relation: Embodied Simulation as Second-Person Perspective on Intersubjectivity. *Philosophical Transactions of the Royal Society of London B: Biological Sciences, 369*(1644). Retrieved from http://rstb.royalsocietypublishing.org/content/369/1644/20130177.short

Gallese, V. (2017). Neoteny and Social Cognition: A Neuroscientific Perspective on Embodiment. In *Embodiment, Enaction, and Culture.* Cambridge, MA: The MIT Press. https://doi.org/10.7551/mitpress/9780262035552.003.0017

Gallese, V., Eagle, M. N., & Migone, P. (2007). Intentional Attunement: Mirror Neurons and the Neural Underpinnings of Interpersonal Relations. *Journal of the American Psychoanalytic Association, 55*(1), 131–175. https://doi.org/10.1177/00030651070550010601

Gallese, V., Keysers, C., & Rizzolatti, G. (2004). A Unifying View of the Basis of Social Cognition. *Trends in Cognitive Sciences, 8*(9), 396–403. https://doi.org/10.1016/j.tics.2004.07.002

Giambra, L. M. (1995). A Laboratory Method for Investigating Influences on Switching Attention to Task-Unrelated Imagery and Thought. *Consciousness and Cognition, 4*(1), 1–21. https://doi.org/10.1006/ccog.1995.1001

Glenberg, A. M. (2010). Embodiment as a Unifying Perspective for Psychology. *Wiley Interdisciplinary Reviews: Cognitive Science, 1*(4), 586–596. https://doi.org/10.1002/wcs.55

Grant, E. R., & Spivey, M. J. (2003). Eye Movements and Problem Solving: Guiding Attention Guides Thought. *Psychological Science, 14*(5), 462–466. https://doi.org/10.1111/1467-9280.02454

Hutto, D. D. (2015). Basic Social Cognition Without Mindreading: Minding Minds Without Attributing Contents. *Synthese, 194*(3), 827–846. https://doi.org/10.1007/s11229-015-0831-0

Jung, W. H., Prehn, K., Fang, Z., Korczykowski, M., Kable, J. W., Rao, H., & Robertson, D. C. (2016). Moral Competence and Brain Connectivity: A Resting-State fMRI Study. *NeuroImage, 141*, 408–415. https://doi. org/10.1016/j.neuroimage.2016.07.045

Keysers, C., & Gazzola, V. (2007). Integrating Simulation and Theory of Mind: From Self to Social Cognition. *Trends in Cognitive Sciences, 11*(5), 194–196. https://doi.org/10.1016/j.tics.2007.02.002

Leech, R., Kamourieh, S., Beckmann, C. F., & Sharp, D. J. (2011). Fractionating the Default Mode Network: Distinct Contributions of the Ventral and Dorsal Posterior Cingulate Cortex to Cognitive Control. *The Journal of Neuroscience: The Official Journal of the Society for Neuroscience, 31*(9), 3217–3224. https://doi.org/10.1523/JNEUROSCI.5626-10.2011

Li, W., Mai, X., & Liu, C. (2014). The Default Mode Network and Social Understanding of Others: What Do Brain Connectivity Studies Tell Us. *Frontiers in Human Neuroscience, 8*(74). https://doi.org/10.3389/fnhum. 2014.00074

Mitchell, J. P., Macrae, C. N., & Banaji, M. R. (2006). Dissociable Medial Prefrontal Contributions to Judgments of Similar and Dissimilar Others. *Neuron, 50*(4), 655–663. https://doi.org/10.1016/j.neuron.2006.03.040

Molnar-Szakacs, I. (2011). From Actions to Empathy and Morality—A Neural Perspective. *Journal of Economic Behavior & Organization, 77*(1), 76–85. https://doi.org/10.1016/J.JEBO.2010.02.019

Molnar-Szakacs, I., & Uddin, L. Q. (2013). Self-Processing and the Default Mode Network: Interactions with the Mirror Neuron System. *Frontiers in Human Neuroscience, 7*(571). https://doi.org/10.3389/FNHUM.2013.00571

Parr, L. A., Waller, B. M., & Fugate, J. (2005). Emotional Communication in Primates: Implications for Neurobiology. *Current Opinion in Neurobiology, 15*(6), 716–720. https://doi.org/10.1016/j.conb.2005.10.017

Pascual, L., Rodrigues, P., & Gallardo-Pujol, D. (2013). How Does Morality Work in the Brain? A Functional and Structural Perspective of Moral Behavior. *Frontiers in Integrative Neuroscience, 7*(65). https://doi.org/10.3389/fnint.2013.00065

Pezzulo, G., Barsalou, L. W., Cangelosi, A., Fischer, M. H., McRae, K., & Spivey, M. J. (2012). Computational Grounded Cognition: A New Alliance Between Grounded Cognition and Computational Modeling. *Frontiers in Psychology, 3*, 612. https://doi.org/10.3389/fpsyg.2012.00612

Railton, P. (1984). Alienation, Consequentialism, and the Demands of Morality. *Philosophy and Public Affairs, 13*(2), 134–171. Retrieved from http://philosophyfaculty.ucsd.edu/faculty/rarneson/Courses/railtonalienationconsequentialism.pdf

Ratcliffe, M. (2013). The Structure of Interpersonal Experience (pp. 221–238). https://doi.org/10.1007/978-3-319-01616-0_12

Rizzolatti, G., & Luppino, G. (2001). The Cortical Motor System. *Neuron, 31*(6), 889–901. Retrieved from http://www.ncbi.nlm.nih.gov/pubmed/11580891

Sandrone, S. (2013). Self Through the Mirror (Neurons) and Default Mode Network: What Neuroscientists Found and What Can Still Be Found There. *Frontiers in Human Neuroscience, 7*(383). https://doi.org/10.3389/fnhum.2013.00383

Schippers, M. B., & Keysers, C. (2011). Mapping the Flow of Information Within the Putative Mirror Neuron System During Gesture Observation. *NeuroImage, 57*(1), 37–44. https://doi.org/10.1016/j.neuroimage.2011.02.018

Singer, T., Seymour, B., O'Doherty, J. P., Stephan, K. E., Dolan, R. J., & Frith, C. D. (2006). Empathic Neural Responses Are Modulated by the Perceived Fairness of Others. *Nature, 439*(7075), 466–469. https://doi.org/10.1038/nature04271

Uddin, L. Q., Iacoboni, M., Lange, C., & Keenan, J. P. (2007). The Self and Social Cognition: The Role of Cortical Midline Structures and Mirror Neurons. *Trends in Cognitive Sciences, 11*(4), 153–157. https://doi.org/10.1016/j.tics.2007.01.001

Umiltà, M. A., Kohler, E., Gallese, V., Fogassi, L., Fadiga, L., Keysers, C., & Rizzolatti, G. (2001). I Know What You Are Doing. *Neuron, 31*(1), 155–165. https://doi.org/10.1016/S0896-6273(01)00337-3

Varela, F. J., Thompson, E., & Rosch, E. (1991). *The Embodied Mind: Cognitive Science and Human Experience.* Cambridge, MA: MIT Press.

Wohlschläger, A., Haggard, P., Gesierich, B., & Prinz, W. (2003). The Perceived Onset Time of Self- and Other-Generated Actions. *Psychological Science, 14*(6), 586–591. https://doi.org/10.1046/j.0956-7976.2003.psci_1469.x

Index[1]

---

[1] Note: Page numbers followed by 'n' refer to notes.

© The Author(s) 2019

M. D. Beni, *Structuring the Self*, New Directions in Philosophy and Cognitive Science,
https://doi.org/10.1007/978-3-030-31102-5

CPI Antony Rowe
Eastbourne, UK
January 10, 2020